Inland Navigation: Environmental Sustainability

Edited by

Bruce L. McCartney, P.E., D

Prepared by the Task Committee on Inland Navigation of the
Waterways Committee of the
Coasts, Oceans, Ports, and Rivers Institute of the
American Society of Civil Engineers

ASCE

COPRI

Published by the American Society of Civil Engineers

Library of Congress Cataloging-in-Publication Data

Names: McCartney, Bruce L., editor. | Coasts, Oceans, Ports and Rivers Institute (American Society of Civil Engineers). Task Committee on Inland Navigation, author.
Title: Inland navigation. Environmental sustainability / prepared by the Task Committee on Inland Navigation of the Waterways Committee of the Coasts, Oceans, Ports, and Rivers Institute of the American Society of Civil Engineers ; edited by Bruce L. McCartney, P.E.
Description: Reston, Virginia : The American Society of Civil Engineers, [2018] | Includes bibliographical references and index.
Identifiers: LCCN 2018043317| ISBN 9780784415092 (softcover : alk. paper) | ISBN 9780784481516 (pdf) | ISBN 9780784481943 (epub)
Subjects: LCSH: Stream channelization–United States. | Stream restoration–United States. | Aquatic ecology–United States. | Inland navigation–Environmental aspects–United States. | Sustainable development–United States.
Classification: LCC TC529 .I554 2018 | DDC 333.91/5160973–dc23
LC record available at https://lccn.loc.gov/2018043317

Published by American Society of Civil Engineers
1801 Alexander Bell Drive
Reston, Virginia 20191-4382
www.asce.org/bookstore | ascelibrary.org

***Errata*: Errata, if any, can be found at https://doi.org/10.1061/9780784415092.**

ISBN 978-0-7844-1509-2 (print)
ISBN 978-0-7844-8151-6 (PDF)
ISBN 978-0-7844-8194-3 (ePub)

Manufactured in the United States of America.

25 24 23 22 21 20 19 1 2 3 4 5

Contents

Acknowledgments

This report is produced by a Task Committee of the Waterways Committee of the Coasts, Oceans, Ports, and Rivers Institute (COPRI), American Society of Civil Engineers.

Members of this Task Committee and Authors of this report are

Bruce L. McCartney, P.E., D.NE, M.ASCE, US Army Corps of Engineers (retired), *Committee Chair and Editor*, author of Chapters 1, 3 through 7, and 11 through 13

John L. Childs, P.E., M.ASCE, Research Civil Engineer, US Army Corps of Engineers, Engineering Research and Development Center, author of Chapter 9 and Appendix C

William H. McAnally, Ph.D, P.E., D.NE, F.ASCE, Research Professor Emeritus, Mississippi State University, author of Chapters 2 and 8 and Appendix B on the Tennessee–Tombigbee Waterway

Thomas J. Pokrefke, P.E., D.NE, LM.ASCE, Consultant, Hydraulic Engineer, author of Chapter 10

Waterways Committee Reviewers are

Michael D. Cox, Dist.D.NE, Aff. M.ASCE, Chief of Operations, Rock Island District, US Army Corps of Engineers

Thomas W. Wakeman III, Eng. Sc.D., F.PIANC, Dist.D.NE., F.ASCE, Research Deputy Director of the Stevens Institute of Technology, Hoboken, NJ

Peer Reviewers are

James L. Martin, Ph.D., P.E., DWRE, F.EWRI, F.ASCE, Professor and Kelly Gene Cook Sr., Chair, Department of Civil and Environmental Engineering, Mississippi State University

Ida Royer, Fish Biologist, Bonneville Project, Portland District, US Army Corps of Engineers

Michael P. Whelan, P.E., D.CE, Assistant Director, Waterfront Engineering, Taylor Engineering, Inc., Jacksonville, FL

The efforts of the aforementioned reviewers are much appreciated. To obtain an expert technical review, three elements are essential:

- Extensive knowledge and experience,

- Time to perform the review, and
- Desire to undertake this effort.

All the reviewers exceeded these requirements.

The authors also extend their appreciation to other contributors:

David C. Gorden, St. Louis District, USACE

Freddie Pinkard, Vicksburg District, USACE

Mike Ott, Portland District, USACE

In addition, the editor very much appreciates Carol McAnally's effort in helping assemble this report.

US Customary to SI (Metric) Units Conversion

Length

US Customary Unit	*Equivalent SI (Metric) Unit*
1 inch (in.)	25.4 millimeters (mm)
1 in.	2.54 centimeters (cm)
1 foot (ft)	30.48 cm
1 yard (yd)	0.91 meter (m)
1 mile (mi)	1.61 kilometer (km)

Area

US Customary Unit	*Equivalent SI (Metric) Unit*
1 square inch (in^2)	6.45 square centimeters (cm^2)
1 square foot (ft^2)	0.09 square meter (m^2)
1 square yard (yd^2)	0.84 m^2
1 square mile (mi^2)	2.59 square kilometer (km^2)
1 acre	0.41 hectare (ha)

Mass (Weight)

US Customary Unit	*Equivalent SI (Metric) Unit*
1 ounce (oz)	28.35 grams (g)
1 pound (lb)	0.45 kilogram (kg)
1 short ton (2,000 lbs)	2,240 tons (1,000 kg)

Source: ANSI (https://blog.ansi.org/2018/06/us-customary-system-conversion-metric/).

Volume

US Customary Unit	*Equivalent SI (Metric) Unit*
1 cubic foot per second (ft^3/s)	0.03 cubic meter per second (m^3/s)
1 foot per second (ft/s)	0.3048 meter per second (m/s)
1 gallon per minute (gpm)	3.79 liters per minute (l/min)
1 gallon per second (gps)	3.79 liters per second (l/s)

Source: Convertworld.com (https://www.convertworld.com/en/volumetric-flow-rate/cubic-feet-per-second.html)

CHAPTER 1
Introduction

1.1 PURPOSE

The purpose of this technical report is to provide an overview of the ecosystem sustainability procedures currently used for US inland waterways. Science and engineering are dynamic fields built on past discoveries and knowledge. This report provides the current state of environmental preservation procedures for inland waterways and will be updated as new concepts are developed.

This report is intended as a reference for those involved with waterway design and operations and as an educational text for the academic community.

1.2 SCOPE

This report presents environmental considerations for construction, operation, and restoration of waterways in the continental United States. Navigable waterways in the United States include both coastal (tidal) and inland waterways that have been used in the past, present, or potential future to move people or commodities. Environmental sustainability of waterways is being pursued by the United States and many other nations. Much of the international work is reported by publications of the International Navigation Association (PIANC).

Incorporating foreign country laws, regulations, and environmental issues is beyond the scope of this report. Because the primary designer and operator of US waterways is USACE, the majority of information and examples are taken from USACE project development and operations practice.

This report is limited to inland freshwater waterways that are primarily for barge traffic. The LMR and lower Columbia River accommodate ship traffic but are also included here because the vast majority of flow is freshwater.

Navigation project elements include locks, dams, powerhouses, channels, and river training works.

Topics addressed in the report include hydrology and hydraulics; sedimentation, dredging, and disposal; water quality; habitat (aquatic, terrestrial, wetlands); migratory fish and birds; historic preservation; and restoration and environmental laws directed to waterway design and operation.

The report structure is summarized as follows:

- Chapter 1: Purpose, scope, and background,
- Chapter 2: Definition of environmental sustainability,
- Chapters 3 and 4: History and physical characteristics of major US navigable waterways,
- Chapters 5 and 6: Environmental legislation and application to waterway design and operation,
- Chapter 7: Requirements for environmental impact statement and environmental assessment,
- Chapter 8: Hydraulics and sedimentation processes background,
- Chapter 9: Dredging and disposal background,
- Chapter 10: Sustainability consideration for river training works,
- Chapter 11: Sustainability measures for fisheries,
- Chapter 12: Other sustainability considerations for birds, lands, salinity, and culture,
- Chapter 13: Example projects discussed earlier in the book, and
- Appendixes: Abbreviations, a history of the Tennessee–Tombigbee Waterway Project, and habitat development using dredge material.

The design of new waterways and replacement of locks and dams on existing waterways are presented in ASCE Manual of Practice (MOP) No. 94, *Inland Navigation: Locks, Dams, and Channels* (McCartney et al. 1998). Chapter 14 of MOP 94 provides a brief discussion of the environmental reporting requirements [Environmental Impact Statement (EIS) and Environmental Assessment (EA)] for new construction and rehabilitation projects. Also provided are environmental factors to be analyzed, such as change to sedimentation, water quality, and habitat.

Another ASCE MOP (124) presents a more detailed look at the design of channel training works, *Inland Navigation Channel Training Works* (Pokrefke 2012). MOP 124 includes a chapter called Environment Design, which provides insight on how river training works (dikes, revetments, cutoffs, etc.) can be made environmentally sustainable.

1.3 SUSTAINABILITY

The pursuit of sustainability is based on moral and ethical principles and procedures required by law. The moral and ethical challenge is to provide a viable productive ecosystem for future generations. The sustainability goal is to maintain a natural resource (navigable waterway) in a way that preserves the resource for present and future generations of humans. This requires not the turning back of time to the prehuman habitat but having humans work with nature for the common good. Chapter 2 explores this subject in more detail.

The guidelines to practice of the ASCE Code of Ethics includes the following provision (ASCE 2017):

> Engineers should be committed to improving the environment by adherence to the principles of sustainable development so as to enhance the quality of life of the general public.

Sustainability accomplishment is further defined by ASCE Policy 418 (ASCE 2016):

> The American Society of Civil Engineers (ASCE) defines sustainability as a set of economic, environmental and social conditions … in which all of society has the capacity and opportunity to maintain and improve its quality of life indefinitely without degrading the quantity, quality or the availability of economic, environmental and social resources. Sustainable development is the application of these resources to enhance the safety, welfare, and quality of life for all of society.
>
> ASCE recognizes the leadership role of engineers in sustainable development, and their responsibility to provide quality and innovation in addressing the challenges of sustainability. The ASCE Code of Ethics requires civil engineers to strive to comply with the principles of sustainable development in the performance of their professional duties. ASCE will work on a global scale to promote public recognition and understanding of the needs and opportunities for sustainable development.

Sustainability and other ethical standards related to navigation can be found in MOP 116, *Navigation Engineering Practice and Ethical Standards* (McAnally et al. 2009).

US laws requiring environmental preservation of our waterway ecosystem began with passage of the Fish and Wildlife Coordination Act of 1946. Since then, many more laws have been enacted to preserve, protect, and create sustainable waterways. Chapter 5 provides a list of environmental legislation in place at present.

1.4 PRACTICE

ASCE's Coasts, Oceans, Ports, and Rivers Institute (COPRI) defines *sustainability* as a set of environmental, economic, and social conditions applicable to engineering activities in coastal, ocean, port, waterway, riverine, and wetland settings in which all of society has the capacity and the opportunity to maintain and improve its quality of life indefinitely without degrading the quantity, quality, or availability of natural, economic, and social resources. The sustainability principle is discussed further in Chapter 2.

The general approach to developing an environmentally sustainable waterway includes

- Identifying baseline environmental conditions,
- Designing and constructing projects that minimize adverse ecosystem impacts, and
- Providing mitigation for unavoidable impacts.

A common mitigation practice is to replace habitats that are lost due to the navigation project with equivalent acreages. For example, lands that are inundated because of construction of locks and dams may be replaced by converting nearby farmland to terrestrial habitat. This is accomplished by the purchase of land or an easement of the replacement land. Following acquisition of the replacement land, long-term management is needed to ensure that proper habitat is maintained.

Waters that are cut off from the natural waterway may need to be reconnected to the main channel. These lands include river oxbows and seasonal floodplain areas that are cut off from the river by channel training works or bendways shortened by cutoffs.

Consideration of environmental impacts is an essential element in the design of any project, including navigation projects. This philosophy follows:

> The design for a modern inland waterway is to provide a navigation project that is safe, efficient, reliable, cost effective, and sustainable. Safety, efficiency, and reliability should be achieved before cost is optimized (McCartney 1986, p. 647).

1.5 NAVIGATION ENGINEERING

Navigation engineering (NE) is a specialty in the field of civil engineering. The Academy of Coastal, Ocean, Port and Navigation Engineers (ACOPNE) describes NE as follows (ACOPNE 2015):

> Navigation engineering involves the life-cycle planning, design, construction, operation and maintenance of safe, secure, reliable, efficient, and environmentally sustainable navigable waterways (channels, structures, and support systems) used to move people and goods by waterborne vessels. Navigation engineering involves civil engineering, hydrology, hydraulics, surveying, geodesy, geography, and cartography. Practice may include research and consulting on navigation of inland waterways, navigation for deep-draft vessels, navigation locks and dams, collection of bathymetric data, and planning and layout of navigation projects.

ACOPNE was founded in October 2009 and provides board certification as a Diplomate in one or more of the four disciplines. The Diplomate designation is a voluntary, postlicensure credential that provides recognition of advanced expertise in a technical specialty. The Navigation Engineering Diplomate designation is D.NE.

The origin of the NE concept was an ASCE Waterways Committee initiative. In the late 1980s, the Waterways Committee saw the need for consolidating the piecemeal information and experience in design and operation of navigation projects.

A series of MOPs were selected as the best way to consolidate and preserve navigation project information. This report joins the following ASCE MOPs that resulted from this Waterways Committee initiative:

- No. 94, *Inland Navigation: Locks, Dams, and Channels*, 1998,
- No. 107, *Ship Channel Design and Operation*, 2005,
- No. 116, *Navigation Engineering Practice and Ethical Standards*, 2009, and
- No. 124, *Inland Navigation Channel Training Works*, 2012.

The first reference to the NE concept was in MOP 94:

> The main emphasis of this manual is design of navigation locks, dams, and channels. Other subjects presented in less detail are environmental design, economic analysis, construction, operation and maintenance. It is the hope of this task committee that these related subjects with less coverage could be developed into complementary manuals to provide an in-depth library of publications to capture the complex and diverse subject of inland navigation. (McCartney 1998, pp. 2–3).

The term *navigation engineering* was introduced in MOP 107:

> The ultimate goal of these (MOPs 50, 94, and 107) and future publications on the navigation system is to provide a body of technical literature for development of a "Navigation Engineering" specialty in the Civil Engineering profession. (McCartney 2005, p. 7).

The acceptance of the NE specialty was achieved with the creation of ASCE's Academy of Coastal, Ports, Ocean, and Navigation Engineering (ACOPNE).

1.6 VALUE TO THE NATION

The nation's inland and intracoastal waterway system carries nearly one-sixth of the cargo moved between cities in the United States. The Mississippi River and its tributaries and the Gulf Intracoastal Waterway connect Gulf Coast ports, such as Mobile, New Orleans, Baton Rouge, Houston, and Corpus Christi, with major inland ports, including Memphis, Saint Louis, Chicago, Minneapolis, Cincinnati, and Pittsburgh. The Mississippi River from Baton Rouge to the Gulf of Mexico allows ocean shipping to connect with the barge traffic, thereby making this segment vital to both the domestic and foreign trade of the United States. In the Pacific Northwest, the Columbia–Snake River system allows navigation for 465 mi inland to Lewiston, Idaho.

Barges are particularly well suited for the movement of large quantities of bulk commodities, such as petroleum, coal, grain, and raw materials at a relatively low cost, for example:

- Coal is the largest commodity by volume moving on our inland waterways. America's utility industry depends on the nation's rivers for more than 20% of the coal it consumes to produce the electricity we depend on to run our homes, offices, and industries. Coal accounted for 30.4% in 2016. In 2017, coal accounted for 30% of our nation's electricity production.

- About 22% of domestic petroleum is shipped by water.
- More than 60% of farm exports move on inland waterways such as the lower Mississippi River or lower Columbia River to downstream ports such as New Orleans for shipment overseas. Nearly 80 million tons of grain is moved by barge annually.
- Other major commodities including metal ores; stone, sand, and gravel used in construction; chemicals, including fertilizers; steel and cement; and many other products.

Inland navigation is a key element of state and local government economic development and job creation efforts and is essential to maintaining our nation's economic competitiveness and national security.

Inland and intracoastal waterways directly serve 38 states throughout the nation's heartland, Atlantic seaboard, Gulf Coast, and Pacific Northwest. The shippers and consumers in these states relied on the inland waterways to move more than 622 million tons of cargo in 2007. All domestic waterborne commerce (inland, coastal, and Great Lakes) amounted to more than 1 billion tons with a value of more than $380 billion in 2007. These goods were shipped from 40 states.

- Texas, Louisiana, and Alaska each ship more than $20 billion worth of cargo annually.
- Illinois, New York, California, and Washington each ship between $10 billion and $20 billion yearly.
- Another eight states ship between $5 billion and $10 billion per year.

Sending these goods by water is $11 cheaper per ton compared to other forms of transportation (such as trucks or trains), according to research by the Tennessee Valley Authority. That translates into nearly $7 billion in annual transportation savings for America's economy.

Another, less visible environmental benefit of navigation projects is that they help limit air pollution emissions by enabling tows with many barges to move cargo long distances on considerably less fuel than trains or trucks would need to move the same amount of cargo the same distance. The ability to move much more cargo per shipment by water than is possible by truck or train means that barge transport is both fuel-efficient and environmentally advantageous. On average, a gallon of fuel allows one ton of cargo to be shipped 155 mi by truck, 436 mi by rail, and 576 mi by barge. America's inland waterways allow America to realize tremendous savings in fuel consumption and reduced air emissions from fuel combustion (USACE 2009b).

The US inland waterway system is of great importance for national defense. In times of conflict and crisis, this system has been used to move troops and support materials safely and efficiently. Support of the nation's defense by navigation projects ranges from carrying surge movements of industrial and energy commodities to moving FEMA's stockpiles of strategic commodities. Inland waterways support military preparedness and mobilization installations, fuel deliveries, ordnance works, arsenals, ammunition plants, and depots.

Waterways are critical assets in effective defense industry mobilization and to US defense. The success of a nation in military conflict depends on material production, transportation of materials for that production, and final delivery. A major defense mobilization requirement would induce sharp increases in waterborne traffic of strategic materials such as primary metal products, ores, energy commodities, and chemicals. The major sources of supply and production of these materials are accessible by the national waterways, which ensure secure and efficient support of all types of military operations. Inland waterways serve as primary routes for the movement of products, war material and supplies, oversized machinery, and equipment of strategic national importance (USACE 2009b).

1.7 AUTHORIZATION AND FUNDING FOR NAVIGATION PROJECTS

1.7.1 Authorization

All work on and in the federal waterways in the United States is controlled by Congress. Congress authorizes the study of new waterways and replacement of major components of this system, such as navigation locks. Once a project is authorized, it must be funded by Congress on a yearly basis. Funding is achieved by the authorized budget of the agency charged with construction of the project. The operation and maintenance of navigation projects are also funded annually by Congress.

Construction of a major waterways project takes 5 to 10 years. Some recent major projects include the Tennessee–Tombigbee Waterway, the Red River Waterway (J. Bennett Johnston Waterway), and the Bonneville lock replacement.

Port navigation facilities are funded and built by federal agencies (such as the US Navy), state and local governments, and private concerns. They are usually, but not always, served by a federal navigation channel. On the Tennessee–Tombigbee Waterway, there are more than a dozen such ports. Some ports are maintained by the Corps of Engineers (COE) by direction of the Congress and the president.

1.7.2 Deauthorization

Congress also has the ability to deauthorize projects after original authorization. This deauthorization can occur before construction or after construction has started. An example of deauthorization of a project that was under way is the Cross Florida Barge Canal. The Cross Florida Barge Canal project was intended to move vessels from the Atlantic Ocean to the Gulf of Mexico across the central part of Florida. The project consisted of a 107 mi channel and five locks. The chronology of the authorization to deauthorization follows:

- 1935: Construction began under the Emergency Relief Appropriations Act of 1935. Work was suspended in 1936 when funds were depleted.
- 1942: Congress authorized construction.
- 1964: COE began construction.

- 1969: Suit filed to halt construction (environmental concerns).
- 1971: President signed an executive order to halt construction (project about one-third completed, $74 million spent).
- 1990: Congress passed a bill deauthorizing the project, and the president signed. This led to the creation of a state greenway and conservation area administered by the State of Florida.

1.8 NAVIGABLE WATERWAYS

Navigable waters are considered to be any water body (river or lakes) that has the past, present, or potential in the future to be used for commerce of people or commodities.

References

ACOPNE (Academy of Coastal, Ocean, Port, and Navigation Engineers). 2015. *Navigation engineering body of knowledge*. Accessed November 10, 2015. https://www.acopne.org/certification-certification-exam/navigation-body-knowledge-bok.

ASCE. 2016. *Policy statement 418: The role of the civil engineer in sustainable development*. Accessed August 21, 2017. https://www.asce.org/issues-and-advocacy/public-policy/policy-statement-418—the-role-of-the-civil-engineer-in-sustainable-development/.

ASCE. 2017. *Code of ethics*. Reston, VA: ASCE. Accessed August 21, 2017. https://www.asce.org/code-of-ethics/.

McAnally, W. H., B. L. McCartney, C. Calhoun Jr., M. D. Cox, and T. J. Pokrefke, eds. 2009. *Navigation engineering practice and ethical standards: Manuals and reports on engineering practice no. 116*. Reston, VA: ASCE.

McCartney, B. L. 1986. "Inland waterway project design." *J. Waterway Port Coastal Ocean Eng.* **112** (6): 645–657.

McCartney, B. L., L. L. Ebner, L. Hales, and E. E. Nelson, eds. 2005. *Design and operation: Manuals and reports on engineering practice no. 107*. Reston, VA: ASCE.

McCartney, B. L., J. G. B. K. Lee, M. Lindgren, and F. Neilson, eds. 1998. *Inland navigation: Locks, dams, and channels: Manuals and reports on engineering practice no. 94*. Reston, VA: ASCE.

Pokrefke, T. J., B. L. McCartney, M. D. Cox, S. W. Ellis, D. C. Gordon, W. H. McAnally, and F. Pinkard. 2014. *Inland navigation: Channel training works. Manuals and reports on engineering practice no. 124*. Reston, VA: ASCE.

USACE. 2009a. *Implementation guidelines for Section 2039 of the water resources development act of 2007 (WRDA 2007)—Monitoring ecosystem restoration*. Accessed July 9, 2016. http://planning.usace.army.mil/toolbox/library/WRDA/wrda07Sec2039a.pdf.

USACE. 2009b. *Inland waterway navigation, value to the nation*. Institute of Water Resources. Accessed February 2, 2016. https://www.iwr.usace.army.mil/Portals/70/docs/VTN/VTNInlandNavBro_loresprd.pdf.

CHAPTER 2
Environmental Sustainability

2.1 EVOLUTION OF THE CONCEPT

Waterway design and operation principles in the United States have evolved from a casual disregard of environmental effects prior to the 1950s to a "do no harm" philosophy in the 1960s and 1970s and more recently to the principles of sustainability. ASCE (1998) defines sustainable water resources: "Sustainable water resource systems are those designed and managed to meet the needs of people living in the future as well as those of those of us living today."

The earliest navigation projects were designed and constructed with minimal regard for depletion or destruction of natural resources. Projects were often of too modest a scale for large environmental impacts, and science was inadequate to define any but the most catastrophic effects. Whatever impact they had on their surroundings was ignored unless the project became threatened. Resources seemed abundant compared to their consumption.

The earliest US environmental law relating to navigation—the 1899 US Rivers and Harbors Act (USC 1899)—was designed to protect navigation, not the environment. To protect shipping, the act banned dumping of refuse or fill into waterways. Only in the 1960s was the prohibition on dumping of "refuse" interpreted as preventing pollution.

The dawn of modern environmental consciousness is often linked to two events—the 1962 publication of Rachel Carson's *Silent Spring* (Carson 1962) and the first Earth Day in 1970. Carson's book forecast the loss of thousands of wild species and harm to humans if widespread, indiscriminate use of pesticides continued unabated. Despite furious criticism, accumulating evidence proved Carson right,[1] and public opinion shifted to a majority emphasis on environmental protection. In 1969, Senator Gaylord Nelson called for a nationwide grassroots demonstration on behalf of the environment to be held on April 22, 1970. More than 20 million people responded. Despite opposition to the concept of environmental protection and industrial and political skepticism about the threat of human activities to the

[1]In recent years, a revisionist movement has pointed to a resurgence of malaria in some countries as a sign that DDT should not have been banned. The assertion fails to note that DDT was banned only in North America and some countries in Europe and that malarial resurgence has been tied to DDT-resistant mosquitos, not the absence of DDT.

environment, public support for environmental protection grew, and Congress passed laws designed to reduce environmental damage, as described in Chapter 5.

The new environmental laws and regulations gradually changed the practice of water resources engineering. At first, designers worked primarily to ensure that no laws or regulations were violated while designing for functional and cost efficiency, as evidenced by design guidance wording such as "while fully complying with all applicable environmental laws." Gradually the design emphasis changed to reflect avoidance of environmental harm, with guidance using phrases such as "minimize or eliminate adverse effects to the environment" (USACE 1980). More recently the emphasis has evolved toward natural resources stewardship and the goal of sustainable development. For example, the Tennessee Valley Authority adopted as one policy, "Practice responsible environmental stewardship of the Valley's natural resources," although it still retained the older mind-set with another policy of complying with environmental laws and regulations. COE has established a set of environmental operating principles to ensure that its missions include totally integrated sustainable environmental practices (USACE 2017):

- Foster sustainability as a way of life throughout the organization;
- Proactively consider environmental consequences of all Corps activities and act accordingly;
- Create mutually supporting economic and environmentally sustainable solutions;
- Continue to meet our corporate responsibility and accountability under the law for activities undertaken by the Corps, which may impact human and natural environments;
- Consider the environment in employing a risk management and systems approach throughout life cycles of projects and programs;
- Leverage scientific, economic, and social knowledge to understand the environmental context and effects of Corps actions in a collaborative manner; and
- Employ an open, transparent process that respects views of individuals and groups interested in Corps activities.

These principles must be implemented through COE's regulations and manuals to become embedded in practice and achieve a standing comparable to economic issues.

In addition to laws, regulations, and policies, the principles of sustainable development have now been incorporated into codes of ethics. For example, the ASCE Code of Ethics (ASCE 2017) include the following provision:

> Engineers should be committed to improving the environment by adherence to the principles of sustainable development so as to enhance the quality of life of the general public.

Board certification (Diplomate) in navigation engineering by the ACOPNE requires adherence to the ASCE Code of Ethics as cited. "Demonstrated competence in sustainable engineering with respect to navigation projects is part of the ACOPNE required body of knowledge for navigation engineering" (ACOPNE 2015).

Despite these and similar statements in other engineering codes of ethics, sustainability remains as controversial as basic environmental protection was in the 1970s. A frequent criticism is that true sustainability is idealistic and impossible—any use of resources is bound to decrease the amount available to future generations. However, that criticism is no more valid than saying that we need not design and build for safety, because perfect safety is never achieved. Absolute sustainability can be an ideal goal that is balanced against other goals, like economic development, or sustainability can become a design criterion if properly defined.

2.2 DEFINITION AND POLICIES

The ASCE definition given at the start of this chapter echoes ASCE Policy 418 (ASCE 2016):

> The American Society of Civil Engineers (ASCE) defines sustainability as a set of economic, environmental and social conditions in which all of society has the capacity and opportunity to maintain and improve its quality of life indefinitely without degrading the quantity, quality or the availability of economic, environmental and social resources. Sustainable development is the application of these resources to enhance the safety, welfare, and quality of life for all of society.
>
> ASCE recognizes the leadership role of engineers in sustainable development, and their responsibility to provide quality and innovation in addressing the challenges of sustainability. The ASCE Code of Ethics requires civil engineers to strive to comply with the principles of sustainable development in the performance of their professional duties. ASCE will work on a global scale to promote public recognition and understanding of the needs and opportunities for sustainable development.

Some authors separate sustainability into environmental and economic components; however, the preceding ASCE definition makes no distinction between the two and includes both within the concept and its application.

Waterborne transportation is widely understood to be the most economical form of transport. Less well known is that it has less impact on the air and water quality and natural habitat than equivalent highway, railway, pipeline, or air transport. For example, inland water transport of freight consumes much less oil (thus producing lower emissions) per ton-mile of transport than highway transport and somewhat less than rail transport (Casavant 2000). Although reliable comparative metrics for other forms of resource consumption and degradation are unavailable, waterborne transport may also be more environmentally friendly in terms of habitat fragmentation, water pollution, and habitat destruction. Responsible decision making on transportation policy and investment requires true intermodal performance metrics, including those for sustainability. An ASCE/UNESCO project (ASCE 1998) has offered suggestions for measuring sustainability that can be combined with intermodal transportation metrics (McAnally et al. 2004) to support decision making in transportation.

Definitions and policies similar to these have been adopted by multiple organizations. Some, such as PIANC, have translated sustainability policies into recommended design processes, and others, such as USACE, have expressed them in both design processes and criteria as discussed.

ASCE, along with the American Public Works Association and the American Council of Engineering Companies, has established the Envision infrastructure sustainability rating system within a new Institute for Sustainable Infrastructure for "horizontal" infrastructure service systems (e.g., water, energy, transportation systems) that are generally built horizontally instead of vertically, like buildings (ISI 2017). Envision has 60 sustainability criteria in five sections: (1) Quality of Life, (2) Leadership, (3) Resource Allocation, (4) Natural World, and (5) Climate and Risk. The Natural World section includes credits, such as the following:

1. SITING
   - 1.1 Preserve Prime Habitat
   - 1.2 Protect Wetlands and Surface Water
   - 1.3 Preserve Prime Farmland
   - 1.4 Avoid Adverse Geology
   - 1.5 Preserve Floodplain Functions
   - 1.6 Avoid Unsuitable Development on Steep Slopes
   - 1.7 Preserve Greenfields
2. LAND AND WATER
   - 2.1 Manage Stormwater
   - 2.2 Reduce Pesticide and Fertilizer Impacts
   - 2.3 Prevent Surface and Groundwater Contamination
3. BIODIVERSITY
   - 3.1 Preserve Species Biodiversity
   - 3.2 Control Invasive Species
   - 3.3 Restore Disturbed Soils
   - 3.4 Maintain Wetland and Surface Water Functions

References

ACOPNE (Academy of Coastal, Ocean, Port and Navigation Engineers). 2015. *Navigation engineering body of knowledge*. Accessed November 10, 2015. https://www.acopne.org/certification-certification-exam/navigation-body-knowledge-bok.

ASCE. 1998. *Sustainability criteria for water resource systems: ASCE task committee on sustainability criteria*. New York: ASCE.

ASCE. 2016. *Policy statement 418: The role of the civil engineer in sustainable development*. Accessed August 21, 2017. https://www.asce.org/issues-and-advocacy/public-policy/policy-statement-418—the-role-of-the-civil-engineer-in-sustainable-development/.

ASCE. 2017. *Code of ethics*. Accessed August 21, 2017. https://www.asce.org/code-of-ethics/.

Carson, R. 1962. *Silent spring*. New York: Houghton-Mifflin.

Casavant, J. 2000. "Inland waterborne transportation—An industry under siege." In *Agricultural transportation challenges of the 21st century*. Washington, DC: US Dept. of Agriculture.

ISI (Institute for Sustainable Infrastructure). 2017. *Envision sustainable infrastructure rating system*. Accessed August 23, 2017. https://sustainableinfrastructure.org/envision/.

McAnally, W. H., Y. Zhang, A. J. Allen, R. O. Bowden, and A. Tan. 2004. *Transportation responses to increased Latin American trade*. Starksville, MS: National Center for International Transportation.

USACE. 1980. *Layout and design of shallow-draft waterways*. EM 1110-2-1611. Washington, DC: USACE.

USACE. 2017. *USACE reinvigorates environmental operating principles*. Accessed August 23, 2017. https://www.usace.army.mil/Missions/Environmental/Environmental-Operating-Principles.

USC (United States Code). 1899. *33 United States Code 403*. Washington, DC: Government Printing Office.

CHAPTER 3
History

3.1 GENERAL

The current inland waterway system in the United States consists of about 25,000 miles of inland waterways. About 12,000 miles are maintained by the federal government, and of the more than 220 navigation locks and dam sites, almost all are federally designed and operated. Figure 3-1 shows the US inland waterway system.

The inland waterways were originally used by Native Americans for trade. The system was all open river without channel modifications. The first attempts at controlling rivers for navigation were the removal of snags and obstructions on the Mississippi and Ohio rivers, a mission given to the USACE from Congress in 1824. In addition to clearing and snagging, the COE was authorized to construct wing dams or dikes to concentrate flow into the main channel. This work provided adequate navigation conditions during wet months but not in the dry seasons.

3.2 OHIO RIVER

In 1910 Congress passed the Rivers and Harbors Act, which authorized construction of a series of locks and dams that would provide a 9 ft navigation depth on the Ohio River. When completed in 1929, the "canalization project" consisted of 51 movable dams (with wooden wickets) and lock chambers measuring 600 × 110 ft. At low water stages, the dams were raised to pool water, thus requiring lockage. At high water the wickets were lowered, allowing vessels to avoid the locks through open river navigation. Wicket dam details can be found in MOP 94 (USACE 2015).

3.3 UPPER MISSISSIPPI RIVER

The first lock on the UMR at Keokue, Illinois [Lock and Dam (L&D) 19] was built in 1913 with a 358 ft length and 110 ft width. A second lock, 400 × 56 ft, was built in 1917 at St. Anthony Falls, Minnesota. A third lock, 320 × 80 ft, was built in 1922 at LeClaire, Illinois (L&D 14).

The second period of lock construction on the Mississippi River was between 1930 and 1940, which added an additional 23 locks between St. Louis, Missouri,

Figure 3-1. US inland waterway system.
Source: USACE (2015).

and Minneapolis, Minnesota, and provided a unified lock system for the entire journey. A third period of lock construction on the UMR started in 1957 and continues today. These locks are 1,200 × 110 ft, to accommodate the modern larger tows.

A chronology of navigation development activities on the UMR and Illinois waterways follows (USACE 1993):

Activity	*Year*
Upper Mississippi River:	
Congress authorizes removal of snags and local obstructions	1824
Congress authorizes 4-1/2 ft channel from mouth of Missouri River to St. Paul	1878
Congress authorizes 6 ft channel	1907
Construction of Meeker Island Dam (first L&D 1)	1913
Construction of L&D 19	1914
Construction of L&D 1	1917
Congress authorizes 9 ft deep, 300 ft wide channel, from St. Louis to Cairo, Illinois	1927
Congress authorizes extension of 9 ft channel to St. Paul, Minnesota, through construction of locks and dams	1930
Construction of 29 locks and dams	1930–1940

(Continued)

Activity	*Year*
Construction of L&D 27	1953
Construction of 1,200 ft chamber at L&D 19	1957
Upper and Lower St. Anthony Falls locks authorized	1937
Lower St. Anthony Falls constructed	1956
Upper St. Anthony Falls Lock constructed	1963
Congress authorizes new dam and single 1,200 ft chamber at L&D 26	1978
Congress authorizes construction of second chamber (600 ft) at L&D 26 (R)	1985
Construction of 1,200 ft chamber at Melvin Price Locks and Dam (formerly L&D 26 (R)	1990
Construction of 600 ft chamber (second lock) at Melvin Price Locks	1994
Major rehabilitation/Maintenance	1986–present
Illinois Waterway:	
Congress authorizes construction of the Illinois and Michigan (I & M) Canal	1822
Construction of I & M Canal completed, 6 ft deep channel	1848
Construction of Chicago Sanitary and Ship Canal by Metropolitan Sanitary	
District (MSD), reversing the flow of the Chicago River to reduce pollution of Lake Michigan and one navigation lock and hydropower dam	1900
Construction of four locks and dams for 7 ft channel below LaSalle	
Two built by state of Illinois	1871
Two built by federal government (all locks 75 × 350 ft)	1873
Start of construction of 9 ft navigation project by the state of Illinois from Utica to Lockport (not completed)	1921
Construction of Cal–Sag Channel, reversing the flow of the Calumet River, and one lock by MSD	1922
Construction of present-day 9 ft navigation system of seven locks and dams	1933–1939
Construction of Thomas J. O'Brien Lock and controlling works (part of Cal–Sag channel modification from 60 to 225 ft channel)	1960
Major rehabilitation/maintenance	1975–present

3.4 COLUMBIA–SNAKE RIVERS

The Columbia–Snake rivers developed a high-lift lock and dam system with hydropower features between 1938 and 1975. Locks are 665 to 683 ft long and 84 ft wide to accommodate a 14 ft draft barge.

3.5 THE FUTURE

Three new waterways have been constructed in the last 50 years:

- Snake River, Idaho: four locks and dams, completed 1975,
- Tennessee–Tombigbee (Tenn–Tom) Waterway, Alabama: 10 locks and dams, completed 1985, and
- Red River Waterway, Louisiana: five locks and dams, completed 1987.

As US locks age (80% are more than 50 years old), they need to be rehabilitated or replaced with a new structure. Replacements are generally near the original lock sites. Some recent replacement locks include

- Melvin Price L&D, Mississippi River, Alton, Illinois—completed 1990,
- Robert Bird Locks, Ohio River, Hogsett, West Virgina—completed 1993, and
- Bonneville Lock, Columbia, Bonneville, Oregon—completed 1993.

New waterways may be built in the future. For example, the Coosa River in Alabama and the Trinity River in Texas have been proposed and studied in the past and could be reanalyzed if the economic benefits improve. The Coosa River project would provide a navigation channel and five locks in existing dams and reservoirs. This waterway would connect the Gulf of Mexico through the Port of Mobile to Rome, Georgia. A navigable barge channel presently exists on the Alabama River from Mobile to Montgomery, Alabama. The Coosa project would extend this channel from Montgomery to Rome, Georgia. The Alabama Power Company built five hydropower dams on the Coosa in the 1960s. The navigation project would build five high-lift locks at these dams. The lock lifts would range from 49 to 130 ft. The latest feasibility study made in the 1980s concluded that the economic benefits did not justify the project.

Since the 1880s, a Trinity River barge canal was envisioned between Trinity Bay near Houston and Dallas, Texas. The original plan called for building 36 locks and dams. Several locks and dams were built before World War I when funding was cut off. The Trinity River navigation project was restarted in the 1960s when Congress approved navigation by barges in 1965. Funding was provided for additional studies and surveys, but growing budget problems (the Vietnam War) halted additional funding in 1973. The project is currently inactive but could be pursued in the future.

Future work on US inland waterways will include replacement of current system components and potential construction of new waterways. This work will require dedicated consideration of environmental impacts to ensure sustainability. Information on all US lock and dam systems can be found in MOP 94.

References

USACE. 1993. *Water resources development in Illinois.* Washington, DC: USACE.

USACE. 2015. *History of the navigation development on the Ohio River.* Louisville, KY: US Army Engineer District. Accessed January 13, 2015. https://www.lrl.usace.army.mil/missions/civilworks/navigation/history.aspx.

CHAPTER 4
Characteristics of US Inland Waterways

4.1 GENERAL

The inland waterway system of the United States is dominated by the Mississippi River and its tributaries. The other major system is the Columbia River. The UMR, Ohio, and Columbia systems are characterized by relatively low sediment loads, stable channels, and lock-and-dam type improvements.

The Missouri River, middle Mississippi River, and LMR are alluvial rivers transporting large quantities of fine sediment. River training works (dikes and revetments) are used to stabilize the rapidly shifting channels. Dredging quantities are much larger for the alluvial systems than the "clear-water" rivers.

A physical/environmental description of some of the major US waterways follows.

4.2 UPPER MISSISSIPPI RIVER

The UMR has many islands, chutes, secondary channels, wetlands, and lakes contiguous to the navigation channel. These relatively shallow, slack water areas provide valuable habitat for fish, waterfowl, and wildlife. Environmental concerns include the effect of navigation traffic, impoundment, dredged material disposal, and revetments on habitat and flora/fauna populations.

4.3 MIDDLE AND LOWER MISSISSIPPI RIVER

South of St. Louis, Missouri, the Mississippi River is an open river with no mainstream impoundments. South of Baton Rouge, Louisiana, the river supports deep draft traffic. Major human impacts on the shallow draft portion of the river are because of the dikes and revetments used to contract and stabilize the channel. Riprap is used for revetments north of Cairo, Illinois, and for dikes throughout the shallow-draft, open river; articulated concrete mattress is the major type of revetment south of Cairo. The placement of dikes and revetments has stabilized the river in mostly a fixed alignment so that it is no longer free to create new backwater areas to replace those lost to sedimentation. However, the dikes and revetments provide habitat valuable to some aquatic species, particularly attachment type macro invertebrates. Other engineering works include levees, dredging, and cutoffs. The river is flanked by flood control levees for most of its length, and dredging quantities are high. However, dredging quantities pale when compared to the total sediment load. A series of bendway cutoffs were constructed for flood control on the LMR in the 1930s.

4.4 ILLINOIS WATERWAY

The Illinois Waterway is composed of the Illinois River and a network of canals linking the river's headwaters with Lake Michigan. Major environmental concerns center on sedimentation in wetlands and floodplain lakes. Extensive aquatic and wetland habitat has been eliminated by levee construction, channelization, and land drainage. Waterfowl and freshwater mussel populations have been adversely affected. Pollution from agricultural and municipal sources has abated some in recent years but continues to be a problem. Presently, the major water quality problems are excessive suspended sediment concentrations, turbidity, and dissolved oxygen depletion. The water project has had beneficial impacts on "moist soil" plants that grow on mudflats but detrimental effects on marsh and aquatic vegetation.

4.5 MISSOURI RIVER

The Missouri River was formerly an extremely wide, braided stream. The combined effects of upstream impoundments and dikes and revetments have drastically reduced the sediment load and converted the braided planform into a series of gentle, stable meanders. These modifications have significantly reduced the total water surface area and the amount of available side channel and slack water habitat. Natural processes that create new side channels and wetlands have been almost eliminated by channel stabilization. Channel degradation and bank erosion are problems in some reaches. Upstream reservoir releases are used to augment low flows and provide navigable depths. The navigation season is limited to eight months due to ice and the availability of water. Levees flank much of the length of the waterway.

4.6 ARKANSAS RIVER

The navigation project on the Arkansas River is also referred to as the McClellan–Kerr Arkansas River Navigation System. The waterway follows the Arkansas River from its upstream origin to Mile 19, where it branches through the Arkansas Post Canal to the White River and then follows the White to the Mississippi River.

Mainstream and tributary impoundments and channel stabilization with dikes and revetments have greatly reduced the sediment load and attendant turbidity. Channel stabilization and sedimentation in dike fields is leading to increases in agricultural lands and terrestrial habitat at the expense of aquatic habitat. Some invasive aquatic plant problems have been experienced in mainstream impoundments Water quality problems exist due to high chloride concentrations, particularly in the extreme upstream reaches.

4.7 OHIO RIVER

The Ohio River drains heavily populated regions with three major metropolitan areas (Pittsburgh, Cincinnati, and Louisville) located on the river. Water quality degradation owing to municipal and industrial discharges and spills is a major concern. In the past 25 years, dissolved oxygen levels have increased while bacteria levels decreased due to improved wastewater treatment. Sport and commercial fisheries have also apparently improved. Present concern centers on toxic pollutants, phenols, bacteria, and low-flow dissolved oxygen concentrations. Acid mine drainage is a problem along some tributaries, most notably the Allegheny River. Several of the navigation dams have favorable impacts on dissolved oxygen levels because they mechanically reaerate oxygen-deficient low flows. Storage reservoirs on the tributaries provide considerable low-flow augmentation; during extreme low flow, as much as half of the flow in the system is provided by reservoir storage.

4.8 MOBILE RIVER AND TRIBUTARIES

This system includes the Mobile, Tombigbee, Black Warrior, and Alabama rivers and the Tenn–Tom Waterway. The Tombigbee and Alabama rivers unite just north of Mobile to form the Mobile River. The Black Warrior is a tributary to the Tombigbee. The Tenn–Tom Waterway links the Tennessee River with the Tombigbee. Traffic levels for the Black Warrior–Tombigbee reach are much higher than the Alabama River. Major engineering works include locks and dams, channel enlargement, and bendway cutoffs. There are some river training dikes on the Alabama River. The channel from Demopolis, Alabama, to Mobile is only 150 to 200 ft wide, limiting effective tow size to six barges. The Tenn–Tom Waterway is divided into three sections: the divide section, which is essentially an extended embayment of Pickwick Reservoir; the canal (or chain of lakes) section; and the river section. The canal section consists of five lock pools just east of the Tombigbee River. The pools lie between levees on the west and natural high ground on the east, creating an irregular lakelike eastern shoreline. Flows are regulated to achieve a desirable distribution between the canal and the Tombigbee tributaries. The river section was constructed by building dams and some 35 cutoffs on the Tombigbee River. Key environmental concerns center on mussels, water quality and habitat quality of the cutoff bendways, surface groundwater interactions, and sediment transport.

A description of the Tenn–Tom Waterway and evolution of the project in response to environmental sustainability is given in Appendix B.

4.9 COLUMBIA–SNAKE RIVERS

The Columbia River system is divided into the following segments:

- Pacific Ocean to Portland, Oregon: 43 ft deep ship channel, 110 mi;
- Portland, Oregon, to Bonneville Dam, Oregon: 17 ft deep barge channel, 35 mi; and
- Bonneville Dam, Oregon, to Lewiston, Idaho: 17 ft deep barge channel (for 14 ft draft barges) and eight high-lift locks, 330 mi.

Water quality is generally excellent compared to other major waterways; however, nitrogen supersaturation impacts on the anadromous fishery are a prime concern. Supersaturation at the eight navigation locks and dams is a problem and is compounded by the fact that the high flow (spill) season coincides with major fish migrations. Other hazards to the migratory fishery include the effects of water temperatures elevated by impoundment; fish passage at lock and dams; and predation by sea lions, birds, and squawfish.

4.10 RED RIVER

The Red River Waterway, also known as the J. Bennett Johnston Waterway, is on a tributary of the Mississippi River in Louisiana. The waterway is 235 mi with five locks and dams. The waterway provides a 200 ft wide by 9 ft deep channel for barge traffic from the Mississippi River to Shreveport, Louisiana. The project was completed in December 1994 at a cost of $2.1 billion. There were 36 channel realignment segments, which shortened the river by about 50 mi. The Red River was a braided river system with shallow depths, multiple channels, oxbows, and a seasonal floodplain. The river had a large sediment load, which is the origin of its name.

4.11 GREAT LAKES

The Great Lakes/St. Lawrence Seaway is the most extensive ship channel and lock system in North America. The navigation system is 2,038 nautical mi and extends from the Atlantic Ocean to Duluth, Minnesota. There are 14 locks at six locations with a 600 ft change in elevation from the Atlantic Ocean to Lake Superior (McCartney et al. 2005).

Over the years, the ecosystem productivity has declined, and an extensive restoration was started in 2010. This restoration effort is discussed in Chapter 13.

References

McCartney, B. L., L. L. Ebner, L. Hales, and E. E. Nelson, eds. 2005. *Design and operation, manuals and reports on engineering practice no. 107.* Reston, VA: ASCE.

USACE. n.d. *Environmental engineering for shallow draft waterways.* Draft engineer manual. Washington, DC: USACE.

CHAPTER 5
Environmental Legislation and Implementation

5.1 GENERAL

Waterways have always been used for the water supply for cities and in many cases for waterborne transportation. Early in our country's history, waterways (rivers) were also used as a dumping place for raw sewage and industrial pollution. As the public and Congress became aware of the degeneration of our environment, measures were put in place for ecosystem protection.

An early preservation effort was the creation of national parks, starting with Yellowstone National Park in 1872. There are currently 407 sites managed by the National Park Service with 59 designated as "national parks."

The next preservation measure was creating the National Forest Service to manage federal land. The US Forest Service was created by Congress with the Land Revision Act of 1891. Currently, the US Forest Service manages 155 national forests containing almost 190 million ac of land.

The federal objective of water and related land resources planning is to contribute to national economic development (NED) consistent with protecting the nation's environment, in accordance with national environmental statutes, applicable executive orders, and other federal planning requirements. Water and related land resources project plans are formulated to alleviate problems and take advantage of opportunities in ways that contribute to study planning objectives and, consequently, to the federal objective. Contributions to NED are increases in the net value of the national output of goods and services, expressed in monetary units, and are the direct net benefits that accrue in the planning area and the rest of the nation. Contributions to NED also include increases in the net value of those goods and services that are marketed and of those that may not be marketed. Protection of the nation's environment is achieved when damage to the environment is eliminated or avoided and important cultural and natural aspects of our nation's heritage are preserved. Various environmental statutes and executive orders assist in ensuring that water resource planning is consistent with protection.

5.2 LEGISLATION

A brief description of major legislation related to environmental protection follows:

Fish and Wildlife Coordination Act of 1946, promotes the preservation and enhancement of fish and wildlife through equal consideration of their habitat needs in conjunction with federal participation in water resource development.

Federal Water Pollution Control Act of 1956 and subsequent amendments, provides for the preservation of water quality through low-flow augmentation.

Fish and Wildlife Coordination Act of 1958, provides that equal consideration should be given to fish and wildlife resources through consideration of their habitat needs in conjunction with federal participation in water resource development. This act also provides authority to modify projects for the benefit of fish and wildlife enhancement.

National Historic Preservation Act of 1966, directs federal agencies to take into account the effects of any undertaking (a federally funded or assisted project) on historic properties (adapted from NCSHPO 2016).

NEPA of 1969, outlines the actions to be taken relative to protecting and enhancing the quality of the human environment. In general, it requires that the impacts to the human environment be evaluated as a project is planned, with the impacts presented in the environmental impact statement. Further, this documentation needs to be coordinated with the public so that its comments are considered as the final project is selected.

Federal Water Pollution Control Act of 1972, also referred to as the Clean Water Act (CWA), established goals to restore and maintain the quality of the nation's waters. The effects of the regulation on system water quality are continuously monitored to ensure that system regulation enhances water quality to the extent reasonably possible.

Endangered Species Act (ESA) of 1973, directs that all federal departments and agencies shall seek to conserve endangered and threatened species and shall use their authorities in furtherance of the purposes of this act. The purposes of this act are to provide a means whereby the ecosystems upon which endangered and threatened species depend may be conserved and to provide a program for the conservation of such endangered and threatened species. Section 7 states that all federal departments and agencies shall, in consultation with and with the assistance of the secretary of the interior/commerce, ensure that any actions authorized, funded, or carried out by them are not likely to jeopardize the continued existence of any endangered or threatened species or result in the destruction or adverse modification of habitat determined by the secretary of interior to be critical unless an exception has been granted by the Endangered Species Committee. The US Fish and Wildlife Service of the Department of the Interior administers consultation procedures.

Archaeological and Historic Preservation Act of 1974, specifically Section 469, provides for the preservation of historic and archaeological data (including relics and specimens) that might be irreparably lost or destroyed as a result of any federal construction project or federally licensed activity program. This applies to any alteration of the terrain caused by construction of dams, relocation of railroads and highways, access roads, and erection of workers' communities.

Archaeological Resources Protection Act of 1979, directs the protection of archaeological resources and sites that are on public lands and Indian lands for the present and future benefit of the American people.

Water Resources Development Act (WRDA) of 1986, specifically Section 906, establishes a comprehensive mitigation policy for water resource projects, including Section 906e, which authorizes the secretary of the army to provide for fish and wildlife mitigation resulting in projects under his or her jurisdiction (USACE 2006a).

WRDA of 1986 as amended by WRDA 1996 and WRDA 1999 (Section 1135), provides for ecosystem restoration through modification to COE structures or operation of COE structures or implementation of restoration features when the construction of COE projects has contributed to degradation of the quality of the environment (USACE 2006b).

WRDA of 1992, as amended (Section 204), provides for protection, restoration, and creation of aquatic and wetland habitats in connection with construction and maintenance dredging of an authorized project (USACE 2006b).

WRDA of 2007–Monitoring Ecosystem Restoration, specifically Section 2039, directs ecosystem restoration projects include a plan for monitoring the success of the restoration work (USACE 2009).

5.3 PROJECT DESIGN

Civil works (COE) studies and projects should be in compliance with all applicable federal environmental statutes and regulations and with applicable state laws and regulations where the federal government has clearly waived sovereign immunity. The NEPA requires federal agencies, including COE, to comply with a process that includes the inventory and assessment of the environmental resources within the study area. NEPA also requires the evaluation and comparison of alternatives to determine the impacts to those ecological, cultural, and aesthetic resources identified and investigated. Involvement by resource agencies and the general public during the study process is also required. The NEPA process should include all measures required for compliance with other applicable environmental statutes, such as the ESA, the Clean Air Act, Clean Water Act, Fish and Wildlife Coordination Act, and Historic Preservation Act, among others (USACE 2000).

The following outlines the implementation of environmental laws related to waterway project development (USACE 2000).

Section 904 of the WRDA of 1986 requires the COE to address the following matters in the formulation and evaluation of alternative plans:

- Enhancing national economic development, including benefits to particular regions that are not transferred from other regions;
- Protecting and restoring the quality of the total environment;
- Protecting the well-being of the people of the United State; and
- Preserving cultural and historical values (USACE 2000, pp. 2–5).

COE publications give a useful description of the environmental considerations for waterway projects. With respect to "protecting the nation's environment," COE has adopted the standard that it "is achieved when damage to the environment is eliminated or avoided and important cultural and natural aspects of our nation's heritage are preserved" (USACE 2000, pp. 2–3).

5.4 ECOSYSTEM RESTORATION

The COE incorporated ecosystem restoration as one of the environmental missions within the civil works program in response to the increasing national emphasis on environmental restoration and preservation. Historically, COE involvement in environmental issues focused on compliance with NEPA requirements related to flood protection, navigation, and other project purposes. The ecosystem restoration mission is carried out in addition to activities related to NEPA compliance. Ecosystem restoration can be considered as a single-purpose project or as a part of multiple-purpose projects along with navigation, flood protection, and other purposes whenever restoration improves the value and function of the ecosystem (USACE 2000).

References

NCSHPO (National Conference of State Historical Preservation Officers). 2016. "National historic preservation act of 1966." Accessed July 9, 2016. https://www.ncshpo.org/nhpa1966.shtml.

USACE. 2000. *Planning guidance notebook*. ER1105-2-100. Washington, DC: USACE.

USACE. 2006a. *Missouri River mainstem reservoir system master water control manual*. Washington, DC: USACE.

USACE. 2006b. "Upper Mississippi River system environmental design handbook." Washington, DC: US Army Engineer District, Rock Island District, IL. Accessed February 2, 2016. http://www.mur.usace.army.mil/EMP/designhandbook.htm.

USACE. 2009. "Implementation guidelines for Section 2039 of the water resources development act of 2007 (WRDA 2007)—Monitoring ecosystem restoration." Accessed July 9, 2016. https://www.mvr.usace.army.mil/Portals/48/docs/Environmental/EMP/EMP_Design_Handbook_August_2006.pdf.

CHAPTER 6

Stages of a Navigation Project Life Cycle

6.1 GENERAL

A navigation project evolves from a conceptual plan to a rehabilitation or abandonment at the end of its life cycle. The various phases of the life cycle are

- Planning,
- Design,
- Construction,
- Operation and maintenance,
- Rehabilitation, and
- Restoration of impacted habitat.

An example of the evolution of an inland navigation project is the Tenn–Tom Waterway, which connects the Tennessee and Tombigbee rivers, which flow south to Mobile Harbor and the Gulf of Mexico. A waterway connecting the Tennessee River to the Gulf was suggested as early as the 1700s. A survey of the route was authorized by Congress in 1875 and 1943. Congress authorized the project in 1946 but delayed funding until 1971. Construction began in 1972 and was completed in December 1984. The waterway carries barge traffic with 234 mi of channel and 10 navigation locks. The environmental mitigation plan called for acquisition of 28,000 ac of floodplain and "intensive management."

The project has been an economic and environmental success. Recreational boating and sport fishing far exceeded expectation, and water quality remained good to excellent. A comprehensive case study of the Tenn–Tom Waterway project is included as Appendix B.

6.2 PLANNING

The planning phase of a navigation project includes the determination of economic benefit, environmental impacts, cost of alternatives, and the recommended plan. The benefits and cost estimates for the recommended plan are the basis for the benefit–cost ratio, which is then a major consideration for investment rationale. The environmental impacts are assessed to determine if a project can be built and, if so, the costs of necessary mitigation, which are included in the benefit–cost ratio.

The results of the planning phase are a feasibility report and an environmental impact statement (EIS). These reports are subject to public and other agency review and, in some cases, independent peer review. Finally, the project and all comments are reviewed by the Corps Civil Works Review Board. If accepted, they are sent to Congress for approval. If approved, the design phase of the project can begin.

The economic benefits analysis is one of the major efforts for this feasibility study phase. Typically, this economic analysis is not visited again as the project progresses.

6.3 DESIGN

The design phase starts with the development of an overall project design that is more detailed than the recommended plan in the feasibility report. This design and cost estimate includes any structures, channel dredging, bank protection, relocations, and environmental features (fish ladders, land acquisition for mitigation of lost habitat, etc.). The environmental features are consistent with recommendations in the EIS. This general overall design is the basis for project funding requests to Congress.

The next step is the detailed design phase, which develops a design for each element of the overall project. These elements can be locks, dams, river training works, and other features. Plans and specifications for each feature are developed from these detailed design reports.

Any significant changes in the project scope require a supplement EIS or environmental assessment (EA).

6.4 CONSTRUCTION

The construction phase starts with the contract award for project construction. The project is often accomplished by several contractors with coordination and adherence to plans and specifications by an independent construction management team. A major deviation from the project plans and specifications may require a supplemental EIS or EA.

6.5 OPERATION

The operations project phase starts with the end of construction and continues for the life of the project. For most waterway projects the design life is 50 years, but with adequate maintenance and major rehabilitation work, the project life can be more than 100 years.

An integral part of the operations activities is periodic inspections of structures (locks, dams, powerhouses, river training works, etc.) to assess their structural integrity. This periodic inspection program has been expanded to include assessment of specific goals of an ecosystem, as related to the navigation project. During this ecosystem assessment, AM is a useful tool to adjust required activities as necessary to achieve goals for many natural resources management projects. AM is a concept being used as a method to evaluate specific goals and adjust activities, as necessary, to meet these goals.

Section 2039 of the Water Resources Development Act of 2007, Monitoring Ecosystem Restoration, directed that a proposed ecosystem restoration project include a monitoring plan to determine the success of the ecosystem restoration. The USACE directed the use of adaptive management (AM) for all new restoration projects. An implementation letter to all COE commands (USACE 2009) provides the following guidance:

> **Contingency Plan (Adaptive Management).** An adaptive management plan (i.e., a contingency plan) will be developed for all ecosystem restoration projects. The adaptive management plan must be appropriately scoped to the scale of the project. If the need for a specified adjustment is anticipated due to high uncertainty in achieving the desired outputs/results, the nature and cost of such actions should be explicitly described in the decision document for the project.
>
> AM is a formalized process in which changes to a waterway project configuration or operation can be made in response to analysis of a monitoring program. The elements of an AM plan follow:

- Identify the preproject ecosystem.
- Establish existing conditions.
- Develop plans to improve/restore ecosystem.
- Input and review from all shareholders (state fish and wildlife agencies, local governments, Indian tribes, historical/archeology, community, etc.).
- Recommend systemwide plan.
- Implement recommendations.
- Conduct a monitoring plan to evaluate environmental health and effectiveness of any recommendation.
- Adjust the plan periodically to achieve the goals and objectives.

The AM concept applied to waterway navigation projects would include the mainstream river and tributaries. An example of the AM process is the Missouri River experiment for variable discharges (pulse) from upstream storage dams. A decline of the Lleast tern (endangered), piping plover (threatened), and pallid sturgeon (endangered) prompted a study to create a sustainable habitat for these birds and fish. A plan was developed to use large short-term water releases from Gavins Point Dam (a storage dam on the upper Missouri River). Following this large discharge (pulse), flow was reduced to what was necessary for navigation downstream. The pulse was intended to mimic the natural spring flood cycle.

The pulse plan was endorsed by state and federal resource agencies and approved by the US Fish and Wildlife Service (USFWS). The plan was implemented and monitored for 5 years. Monitoring results were reviewed by a panel of experts. The experts concluded that the pulse plan did not mitigate the losses to fish and birds. The expert panel suggested that the creation of additional habitat may provide and sustain essential resources for these species. A more detailed description of the pulse plan is in Chapter 11.

Another example of AM is the Columbia River Deepening Project (covered in detail in Chapter 13). The project deepened a 40 ft deep channel to 43 ft from Astoria, Oregon, to Portland, Oregon (110 mi). Dredging started in 2005 and ended in 2010. Development of an AM plan started in 2006. Monitoring of the project continued through 2014. Some of the conclusions follow:

- Changes to the physical parameter as a result of the deepening are minor and occur in proximity to the navigation channel.
- Dredging and disposal were as expected. Annual maintenance dredging and placement volumes will continue to be reported.
- Channel site-slope adjustment occurred as expected. The channel improvements had no significant impact on sediment dynamics.

Dredging did not expose organisms to toxic contaminants. The next testing is required after 10 years (in 2018) for compliance with the Sediment Evaluation framework.

Typically, routine maintenance for structures and maintenance dredging using existing placement sites do not need an EIS or EA. Some major changes in project operations or configurations will probably require an EIS or EA.

As projects age, maintenance can become expensive and reliability may decrease. When this occurs, the choices are (1) abandon the project, (2) do nothing and accept decreased reliability, (3) build a new structure, or (4) undertake a major rehabilitation, which can include components for safety, efficiency, and reliability.

USACE's policy on major rehabilitation (USACE 1996) is that, "A major rehabilitation program shall be implemented and maintained for construction of infrequent, costly structural rehabilitation of major replacement works that are intended to improve reliability or efficiency of a Corps project or principal feature thereof."

Major rehabilitation of navigation structures (locks and dams) will need an EIS or EA. An example of a major rehab project is L&D 27 on the Mississippi River. A summary of the EA for this project is presented in the case studies in Chapter 13.

6.6 RESTORATION

The objective of ecosystem restoration is to restore degraded ecosystem structure, function, and dynamic processes to a less degraded, more natural condition. Indicators of success include the presence of a large variety of native plants and animals, the ability of the area to sustain larger numbers of certain indicator species or more biologically desirable species, and the ability of the restored area to continue to function and produce the desired outputs with a minimum of continuing human intervention.

A wide range of improvements to ecosystem functions can include, but are not limited to, using dredged material to restore wetlands; restoring floodplain function by reconnection of oxbows to the main channel; providing for more natural channel conditions, including restoration of riparian vegetation, pools, and riffles and adding structure; modifying obstructions to fish passage, including dam removal; modifying dams to improve dissolved oxygen levels or temperature downstream; removing drainage structures and/or levees to restore wetland hydrology; and restoring conditions conducive to native aquatic and riparian vegetation (USACE 2000).

Examples of recent navigation restoration projects that are presented more detail in Chapter 13 include

- UMR Restoration Plan;
- Willamette River, Oregon, restoration effort to restore the endangered Oregon chub population;
- LMR restoration of habitat to aid three endangered species; and
- Great Lakes restoration for water quality and habitat.

References

USACE. 1996. *Partners and support (work management policies).* ER 1130-2-500. Washington, DC: USACE.

USACE. 2000. *Planning guidance notebook.* ER 1105-2-100. Washington, DC: USACE.

USACE. 2009. *Implementation guidelines for Section 2039 of the water resources development act of 2007 (WRDA 2007)—Monitoring ecosystem restoration.* Accessed July 9, 2016. http://planning.usace.army.mil/toolbox/library/WRDA/wrda07Sec2039a.pdf.

CHAPTER 7

Environmental Impact Statement and Environmental Assessment

7.1 AUTHORITY

With the passage and implementation of the NEPA of 1969 (PL 91-190), environmental impact assessments of water resource projects under the USACE and other federal agencies assumed a greater level of importance. Previously environmental assessments were controlled by internal regulations and were usually not distributed or reviewed outside the agency; subsequently, NEPA established a broad national policy directing federal agencies to maintain and preserve environmental quality.

Section 102(a)c of NEPA requires all federal agencies and officials to (1) direct their policies, plans, and programs to protect and enhance environmental quality; (2) view their actions in a manner that will encourage productive and enjoyable harmony between man and his environment; (3) promote efforts that will minimize or eliminate adverse effects to the environment and stimulate the health and well-being of human beings; (4) promote the understanding of ecological systems and natural resources important to the nation; (5) use a systematic and interdisciplinary approach that integrates the ecological, social, cultural, and economic factors in planning and decision making; (6) study, develop, and describe alternative actions that will avoid or minimize adverse impacts; and (7) evaluate the short-term and long-term impacts of proposed actions (USACE 1988).

7.2 ENVIRONMENTAL IMPACT STATEMENT

An EIS is needed when a project is expected to have a significant impact on the existing ecosystem. An EIS is normally needed for the following:

- New project construction. Needed with feasibility report and used as a basis for authorization.
- Changes to new project after authorization that would increase size substantially or add additional purposes.

- Major changes to operation or maintenance of completed projects, including restoration projects.

An EIS should include an evaluation of project impacts on

- Aquatic habitat,
- Terrestrial habitat,
- Historic and archeological sites, and
- Endangered species.

A mitigation plan is then needed to address the adverse impacts of the proposal plan. The EIS is reviewed and agreed to by affected resource agencies (state fish and wildlife, historic societies, Native American tribes, etc.). Following an agreement with resource agencies, the EIS is forwarded to the USFWS for concurrence. The approved EIS is included with a report that recommends construction (usually a feasibility report) and forwarded to Congress for authorization. Examples of EISs, such as the Red River Waterway and Columbia River Deepening projects, are presented in Chapter 13.

7.3 CUMULATIVE EFFECTS

Part of an EIS is an evaluation of the ecosystem effects of the proposed project plus those of all previous projects—the cumulative effects. The CEQ regulations (40 CFR §§ 1500-1508) implementing the procedural provisions of the NEPA of 1969, as amended (42 U.S.C. § 4321 et seq.), define cumulative impact (40 CFR § 1508.7) as

> [T]he impact on the environment which results from the incremental impact of the action when added to other past, present, and reasonably foreseeable future actions regardless of what agency (federal or non-federal) or person undertakes such other actions.

Cumulative effects analysis recognizes that the most serious environmental impacts may result from the combination of individually minor effects of multiple actions over time rather than the direct or indirect effects of a particular action (CEQ 1997). Cumulative effects or cumulative impacts analyses began to be conducted in the early 1980s, but only recently have these analyses been examined in terms of their importance, effectiveness, and the challenges in their conduct.

The following is the CEQ framework for conducting cumulative impact assessments (CEQ 1997).

1. Cumulative effects are caused by the aggregate of past, present, and reasonably foreseeable future actions. The effects of a proposed action on a given resource, ecosystem, and human community include the present and future effects added to the effects that have taken place in the past. Such cumulative effects must also be added to effects (past, present, and future) caused by all other actions that affect the same resource.

2. Cumulative effects are the total effect, including both direct and indirect effects, on a given resource, ecosystem, and human community of all actions taken, no matter who (federal, nonfederal, or private) has taken the actions. Individual effects from disparate activities may add up or interact to cause additional effects not apparent when looking at the individual effects one at a time. The additional effects contributed by actions unrelated to the proposed action must be included in the analysis of cumulative effects.

3. Cumulative effects need to be analyzed in terms of the specific resource, ecosystem, and human community being affected. Environmental effects are often evaluated from the perspective of the proposed action. Analyzing cumulative effects requires focusing on the resource, ecosystem, and human community that may be affected and developing an adequate understanding of how the resources are susceptible to effects.

4. It is not practical to analyze the cumulative effects of an action on the universe; the list of environmental effects must focus on those that are truly meaningful. For cumulative effects analysis to help the decision maker and inform interested parties, it must be limited through scoping to effects that can be evaluated meaningfully. The boundaries for evaluating cumulative effects should be expanded to the point at which the resource is no longer affected significantly or the effects are no longer of interest to affected parties.

5. Cumulative effects on a given resource, ecosystem, and human community are rarely aligned with political or administrative boundaries. Resources typically are demarcated according to agency responsibilities, county lines, grazing allotments, or other administrative boundaries. Because natural and sociocultural resources are not usually so aligned, each political entity actually manages only a piece of the affected resource or ecosystem. Cumulative effects analysis on natural systems must use natural ecological boundaries, and analysis of human communities must use actual sociocultural boundaries to ensure that all effects are included.

6. Cumulative effects may result from the accumulation of similar effects or the synergistic interaction of different effects. Repeated actions may cause effects to build up through simple addition (more and more of the same type of effect), and the same or different actions may produce effects that interact to produce cumulative effects greater than the sum of the effects.

7. Cumulative effects may last for many years beyond the life of the action that caused the effects. Some actions cause damage lasting far longer than the life of the action itself (e.g., acid mine drainage, radioactive waste contamination, species extinctions). Cumulative effects analysis needs to apply the best science and forecasting techniques to assess potential catastrophic consequences in the future.

8. Each affected resource, ecosystem, and human community must be analyzed in terms of its capacity to accommodate additional effects, based on its own time and space parameters. Analysts tend to think in terms of how the resource, ecosystem, and human community will be modified given the action's development needs. The most effective cumulative effects analysis focuses on what is needed to ensure long-term productivity or sustainability of the resource. (CEQ 1997, Table 1–2, p. 8)

7.4 ENVIRONMENTAL ASSESSMENT AND FINDING OF NO SIGNIFICANT IMPACT

An EA is used when there are no significant impacts expected from a recommended action for a navigation project. An EA covers the same elements and coordination as an EIS but is less detailed.

A FONSI is the transmitting mechanism of the EA, where the requesting agency states that no adverse impacts are expected from the proposed action. The FONSI will be a brief summary document.

Actions normally requiring an EA include

- Regulatory actions (permits);
- Continuing authorities programs (Small Navigation Project, Section 107);
- Minor changes to a project construction, operation, or maintenance (e.g., pool levels, new dredge disposal areas, and bank protection modifications);
- Real estate management and disposal actions;
- Real estate grants for right-of-way, which involve only minor disturbance to earth, air, or water; and
- Transfer or grants of land to other federal agencies.

7.5 EXCLUSIONS FROM ENVIRONMENTAL IMPACT STATEMENT OR ENVIRONMENTAL ASSESSMENT

- Emergency situations to prevent or reduce imminent risk to life, health, property, or severe economic loss. When possible, emergency actions considered major in scope may need an EIS after the event.
- Activities at completed projects that carry out the authorized project purposes. Examples include routine operation and maintenance actions; general administration; equipment purchases; custodial actions; erosion control; painting; repair; rehabilitation; replacement of existing structures and facilities such as buildings, roads, levees, groins, and utilities; and installation of new buildings, utilities, or roadways in developed areas.
- Minor maintenance dredging using existing disposal sites.

References

CEQ (Council on Environmental Quality). 1997. *Considering cumulative effects*. Washington, DC: Executive Office of the President.

USACE. 1980a. *Hydraulic design of reservoir works*. EM 1110-2-1602. Washington, DC: USACE.

USACE. 1980b. *Layout and design of shallow-draft waterways*. EM 1110-2-1611. Washington, DC: USACE.

USACE. 1988. *Procedures for implementing NEPA*. ER 200-2-2. Washington, DC: USACE.

CHAPTER 8

Hydraulics and Sedimentation of Rivers

This chapter provides information on hydraulics and sedimentation processes common to inland waterways to provide background information for environmental situations described in subsequent chapters.

8.1 NAVIGABLE INLAND WATERWAYS

8.1.1 Natural Waterways

Natural inland waterways include most rivers and lakes where water is already available, even if not always at a depth necessary for navigation. Navigability of a waterway depends on sufficient depth and width for a given vessel size. Obviously, a channel just deep enough to accommodate a recreational fishing boat will not allow safe passage of a pushboat with a 6 ft draft. Navigability also implies the safe passage of a vessel—with sufficient under-keel clearance to avoid damage to the vessel or the waterway bed, without obstructions, and without currents that excessively impair the vessel's handling or progress.

Some rivers and reaches of rivers are naturally deep enough to allow safe and effective navigation. Others may require the creation of a deeper channel by structural scouring, dredging, or ponding by dams. The Mississippi River exhibits all three types of channel-deepening methods: navigation dams above St. Louis, Missouri, and structures plus dredging of bars over the river's length.

8.1.2 Canals

Canals are often constructed by excavation to extend or connect natural waterways. The Erie Canal was constructed in the nineteenth century to connect the Hudson River at Albany, New York, to Lake Erie at Buffalo, New York. Canals are constructed no wider and no deeper than necessary to accommodate the largest vessels intended for use and may restrict traffic to one direction at a time except for passing zones.

8.1.3 Lakes and Impoundments

The US–Canadian Great Lakes are the prime example of natural waterways consisting of water impounded by natural geographic features and connected by rivers. Many navigation dams, including those on the UMR, are designed as *run-of-the-river* structures not to store substantial amounts of water but to create a milder river slope that covers shallow points along the upstream reach. Exceptions include many dams on the Columbia–Snake system, such as Bonneville Dam, which store water for hydropower and conservation purposes.

8.2 HYDROLOGY, WEATHER, AND CLIMATE

Hydrology is the science of the water cycle—from precipitation to runoff over the land surface and infiltration into the ground, collection into ponds and streams, and evaporation and plant transpiration back into the atmosphere. *Weather* is the short-term behavior of the atmosphere—temperature, wind, and precipitation on a daily time scale. *Climate* is the long-term pattern of weather on a scale of years to decades and longer.

8.2.1 Precipitation to Runoff to Groundwater

Precipitation occurs as rainfall, snow, and condensation of atmospheric moisture. It is usually recorded as inches (per square in.) or millimeters (per square mm) and expressed in daily, monthly, or annual totals. For some purposes, such as calculating runoff rate, it may be expressed as an intensity, such as inches per 1, 4, or 24 h, or an accumulated volume such as acre-feet or cubic meters.

Runoff versus infiltration rates are calculated as functions of the topographic slope, vegetative land cover, and land surface type and condition. Steep, long slopes without vegetation and impermeable surface produce the most runoff. Flat vegetated areas with high permeability and low soil moisture produce the greatest infiltration to the ground. Runoff is usually expressed as a volumetric discharge in cfs (cubic feet per second, ft^3/s) or cms (cubic meters per second, m^3/s). Runoff occurs as shallow sheet flow over the ground, small rivulets in surface rills, and larger rivulets in eroded gullies. Once in a stream or river, the flow rate may be expressed as a speed in fps (feet per second, ft/s) or mps (meters per second, m/s). Water level, or stage, is easier to measure and is expressed as height (feet or meters) above a given datum plane, such as the NAVD of 1988 (NAVD88).

Infiltrated water and air occupy the small gaps among soil grains in the unsaturated groundwater zone. If most or all the air has been replaced by water, the soil is said to be saturated. Groundwater flows very slowly in most circumstances, typically on the order of feet per day, except in unusual situations such as underground cavities. Groundwater can flow out into surface waters if the level of the ground surface dips below the groundwater table. In dry seasons when precipitation is not contributing the stream flow, groundwater flows into streams make up the streams' "base flow" or minimum annual flow.

8.2.2 Ice

Ice and snow hold water in place until melting occurs. Annual spring snowmelt is a major source of spring floods in the northern half of the United States and in Canada. Ice formation on streams and rivers creates a lid below which water flows until spring breakup, when the ice fractures and moves downstream as floes. Floes sometimes create ice jams, which block the flow and cause flooding. Surface ice buildup can also block gates on locks and dams.

8.2.3 Sediment Yield

Sediment consists of mineral particles typically ranging in size from gravel (greater than 2 mm diameter) to clay (less than 0.002 mm) plus organic material such as leaves and other detritus. Precipitation striking the ground and flowing over the surface picks up sediment, transports it downslope, and delivers it to streams. Gully and stream flow carry the sediment and erode the surface below and on the sides, contributing more sediment to the flow. Exposed and disturbed earth such as freshly plowed fields produce more sediment than heavily vegetated areas such as woodlands. Steeper slopes produce more sediment yield than flatter surfaces. Sediment yield is often expressed in weight or mass per year and can be calculated by empirical techniques similar to those for runoff.

8.2.4 Cycles and Trends

River flows exhibit seasonal cycles in concert with the weather. For example, the Mississippi River experiences two periods of higher flow—a spring freshet caused by snowmelt in the upper watershed and a winter freshet caused by rainfall in the middle and lower watershed. Figure 8-1 shows river stage hydrographs (long-term average and 2003) for the Mississippi River at Cape Girardeau, Missouri, displaying a major peak in April, a minor peak in November-December, and low flows in January and October. Many North American rivers experience the lowest flows in October, so hydrologic data are usually recorded in Water Years—October 1 through September 30—and labeled as the calendar year in which they end. Thus, WY2003 began in October 2002 and ended in September 2003.

Wet and dry cycles correlating with global cycles such as El Niño and La Niña add variability to the expected annual hydrograph pattern, as do shorter and longer periodic oscillations of both known and unknown causes. Extended wet and dry cycles produce localized and widespread drought and flooding periods.

World weather records are too short to delineate cycles longer than a few hundred years, but the geologic record illustrates cycles such as glaciations that lasted thousands of years. The last 50 years or so have produced a measurable increase in rainfall intensity with time in North America; this cycle may eventually produce a decrease in intensity or may be a trend correlated with climate change.

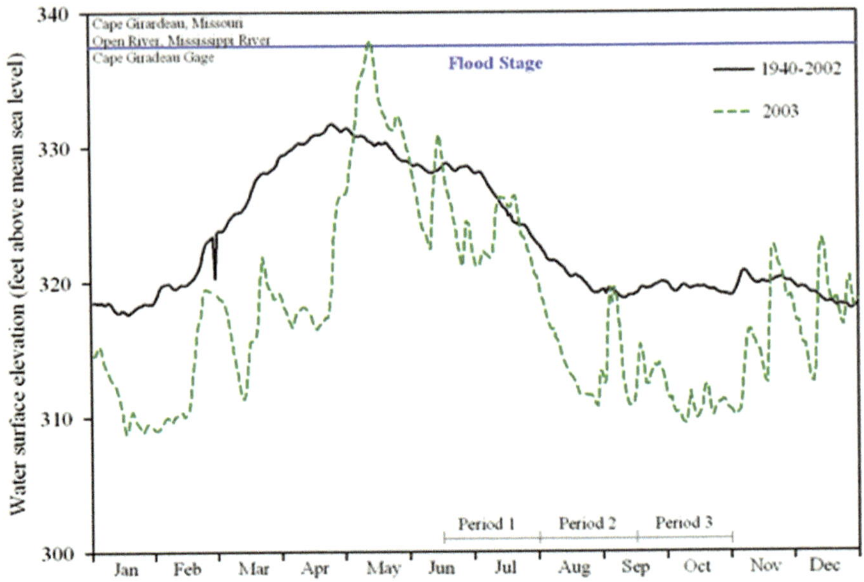

Figure 8-1. Mississippi River example hydrograph.
Source: Courtesy of the National Weather Service.

Global climate change has been unfortunately politicized in the United States, but it is most definitely happening and must be considered in all water resources planning and decision making.

8.2.5 Statistics

Hydrology, weather, and climate are described in statistical terms when used in waterway studies. In addition to the usual measures of maximum, mean, minimum, and standard deviation, measures specific to environmental uses are employed, including extreme low flows, low flows, high flow pulses, small floods, and large floods.

For environmental studies, low flows are commonly specified as the 7Q10 flow (lowest 7-day average flow with a 10-year return interval) or 30Q2 (lowest 30-day average with a 2-year return interval).

Statistical distributions used in hydrologic studies include the normal, log-normal, Gumbel, and log-Pearson Type III probability distributions, and care must be taken to use an appropriate distribution for the application.

8.2.6 Further Information

Additional information on hydrology, weather, and climate can be found in textbooks and manuals, such as

- Ponce (1994),
- Bedient et al. (2012),

- IPCC (2014),
- Clarke (1998), and
- Poff et al. (1997).

8.3 GEOMORPHOLOGY

8.3.1 Landforms

Land topography and texture are affected by hydrology and hydraulics. As previously noted, water runs off steep terrain more rapidly, but if the terrain is erodible, soil runoff will carve gullies that can become streams and concentrate the flow in narrow zones. Sediment eroded from hills and mountains deposits when the ground slope flattens out, creating delta splays. Weather and runoff continually wear down high ground and move its material to lower levels, including filling up valleys, ponds, lakes, and estuaries. Absent tectonics, the process tends to flatten and lower the landscape over time.

8.3.2 River Patterns

Rivers form typical patterns that vary with flow rate, slope of the land, ground conditions, and sediment supply. Steep streams tend to flow in straight lines between inflection points, producing zigzag paths in which smaller streams join in a treelike pattern called *dendritic*. On somewhat flatter slopes with an abundant sediment supply, rivers may become braided, with multiple flow paths among islands. Low slope rivers in alluvium meander, forming loops that grow laterally until a cutoff occurs, leaving behind an oxbow lake typical of the Mississippi River valley.

Rivers form floodplains between confining valley walls as shown in Figure 8-2, filling much of the valley during floods and occupying only a narrow channel during low flow periods. Annual high water may be mostly contained between riverbanks, with less frequent floods spill over those banks, inundating all or part of

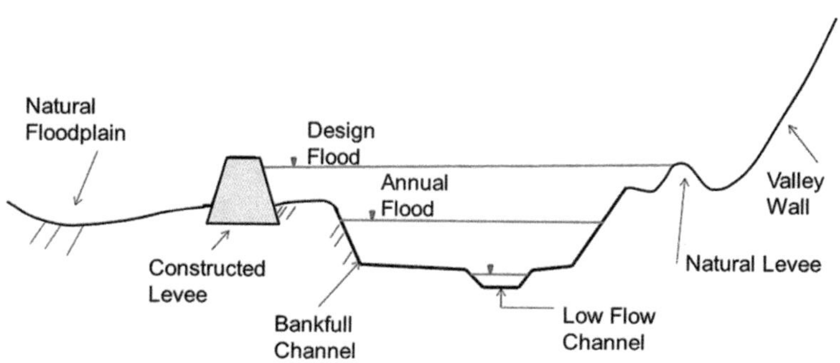

Figure 8-2. Typical floodplain diagram.

the floodplain and depositing sediment in the overbank area before receding to within-banks flow after the flood has passed. This pattern of low flow channels and high flow channels creates natural levees atop banklines and is mimicked by constructed levees, which are designed to contain floods with a specific frequency of occurrence. For example, river levees may be built to contain a 100-year flood (1% probability of occurring in any year) but allow a 200-year (0.5% probability) or greater flood to overtop the levee.

Backwaters are ponded areas adjacent to a river. They may consist of low areas in the floodplain, such as marshes, oxbow lakes, or tributary streams and their floodplains. During main stem river high stages, tributary flow, rainfall, and groundwater seepage fill the backwater.

8.3.3 Further Information

Additional information on geomorphology can be found in publications, such as Julian (2002) and Anderson and Anderson (2010).

8.4 HYDRAULICS

Hydraulics within the field of NE concerns the motion, forces, and effects of water flowing in pipes, surface waters, and under the ground. Although the term *hydraulics* is also used to describe the pressurized flow of oil for controlling lock and dam gates, valves, and other equipment, that use is not examined here.

Water motion is typically described in terms of water surface elevation and velocity. Velocity, speed, and direction include the rate at which the water travels and celerity of waves in the water. Volumetric discharge expresses how much water is moving in cfs or cms, averaged over some period, ranging from minutes to days. Materials transported by the flow in solution or suspension move at about the same speed as the water. Forces exerted by the water on the bottom and banks of its channel include normal pressures—mainly hydrostatic—and shear or drag forces. These forces can usefully hold navigation lock gates closed, disastrously burst levees and walls, and erode bed and banks.

Water motion changes in time, termed *unsteady* flow, and in location, termed *nonuniform* flow. These odd terms (expressing the nature of flow in terms of what it is not) are a historical artifact. The earliest hydraulic engineers could write solvable equations only in terms of steady, uniform flow, so those were the standard solutions by which Egyptians estimated River Nile discharges, Chinese calculated Dujiangyan flood stages, and Romans designed aqueducts. Their steady flow calculations were close enough that they could approximate the actual unsteady, nonuniform flows. Now that hydraulic engineers have more powerful computational methods and machines, they can write and solve time- and space-varying equations routinely, but the nomenclature has been set.

8.4.1 Stage and Slope

Water level is most often expressed as the height of water surface (stage) above some datum and is specific to time and location. For example, Figure 8-3 illustrates stages at two Mississippi River stations over time, displaying the rise and fall of the stage. Datum plane may be gage datum—an arbitrary level set for that gage only—or a general datum plane used over a wide area. The NAVD88 is the official datum plane for most stage data, but there are a few local datum planes, such as the Columbia River Datum, and many legacy datum planes, such as Memphis Datum and Mean Sea Level.

8.4.2 Velocity and Discharge

Flow rate is expressed as a *velocity* (direction and distance traveled by water per unit time) and as a *discharge* (quantity, usually volume, per unit time). Velocity can be defined at a point or as an average over some spatial extent, such as a cross-section or a vertical line. A velocity average over about 1 min is typically used because turbulence introduces wide fluctuations on the order of seconds; however, for analysis purposes, the averaging period can be at daily, monthly, or any useful

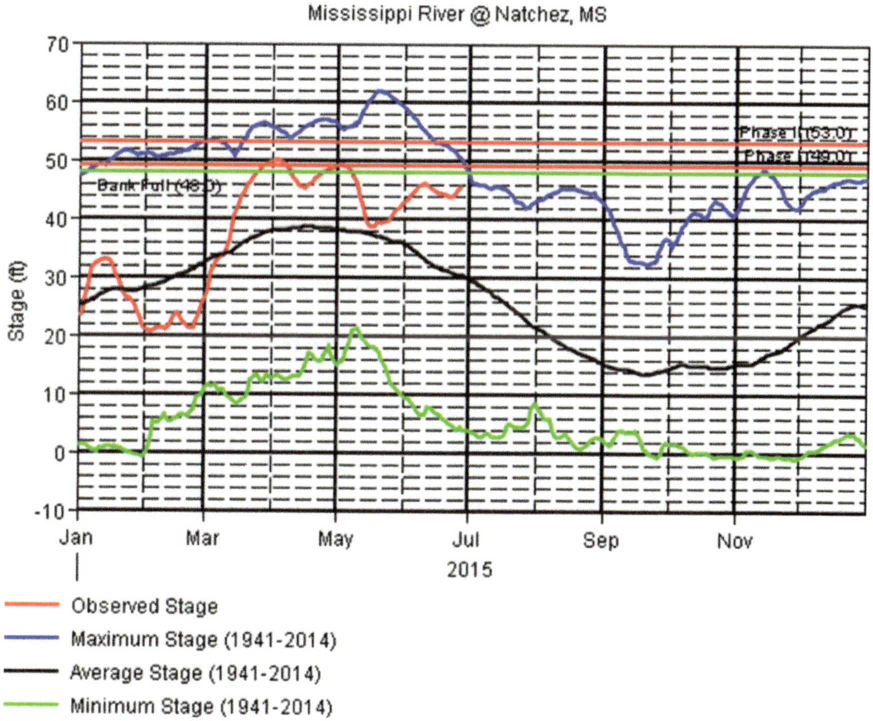

Figure 8-3. Example river hydrographs.
Source: Courtesy USACE.

time scale. Steady flow velocity is proportional to water surface slope and inversely proportional to resistance from the bed and banks of the waterway and can be expressed by the widely used Manning and Chézy equations, among others. In addition to defining how fast the water moves, it is also the speed at which a dissolved constituent moves with the water and approximately the speed at which a suspended particle travels. Thus, a material in solution will travel downstream at the local speed of the water. Materials in transport are also subject to *diffusion*, or the movement by molecular and turbulent eddy motions from zones of higher concentration to zones of lower concentration, and *dispersion*, or the smearing of calculated or measured concentrations by averaging over space or time. The velocity of suspended particles, such as sediment grains, approaches that of the water, but their average speed will be affected by settling, drag, and any time spent on the bed.

Velocity is typically lowest near the bed and banks of a waterway and highest just below the surface in the central region, although the thread of highest speed meanders horizontally and vertically within a flowing stream. The velocity profile in the boundary layer—the zone close to solid surfaces—is described by logarithmic or power law equations that go to zero at or near the solid surface.

Celerity is the speed of a wave at the boundary between two fluids—air and water at the surface or less dense and more dense water at an interface in the water column. Wave celerity is a function of water depth and wave period and is not the same as the flow velocity. (The distinction between wave celerity and flow velocity is equivalent to that between the speed of a sound wave in air and the wind velocity.) Waves caused by wind or lock and dam operations can even move upstream in opposition to the current. Flood waves are a special case of combined disturbance and discharge, and they may be faster or slower than the flow velocity depending on the channel cross-sectional shape. Figure 8-4 illustrates a flood wave propagating down a river valley.

Discharge, typically expressed as cubic feet or meters per second, gallons or liters per day, or acre-ft (1 ac × 1 ft) per year, applies to an entire cross-section. The exception is unit discharge, which is the volumetric flow rate per unit width, such as cubic feet per second per foot of width. If the letter Q is used for volumetric discharge, then the lowercase q denotes unit discharge. Discharge values by themselves are significant for flooding and water supply issues; when combined with information such as slope and cross-sectional area, velocities can be calculated; with constituent concentrations, mass and volume transport rates for pollutants, nutrients, and sediment can be estimated.

8.4.3 Floods and Low Flows

Natural river and stream discharges and the water levels associated with them are

- Unsteady and nonuniform—variable in time and space,
- Stochastic random variables—values vary with an element of chance,
- Continuous—there are an infinite number of possible values over any range, and

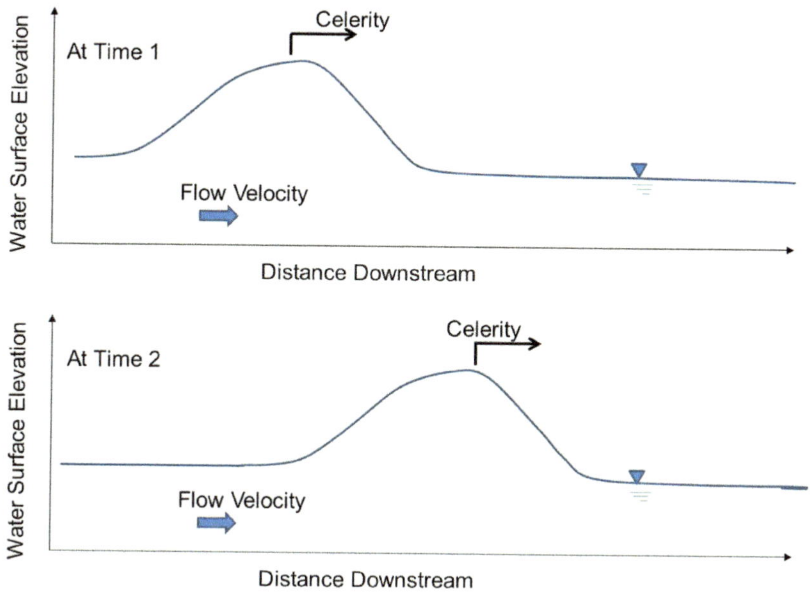

Figure 8-4. Flood wave schematic.

- Distributed randomly—the likelihood of a given value being exceeded can be described by a probability density function (pdf).

Figure 8-5 illustrates probability distribution of flows, with a Mississippi River distribution. Randomness in water data is driven mainly by randomness in precipitation with contributions of a host of other influences, such as ground cover, evapotranspiration, and infiltration.

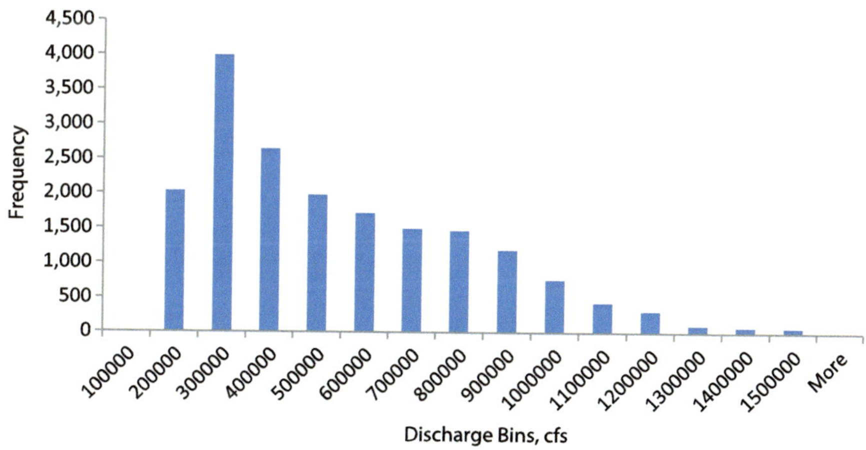

Figure 8-5. Example hydrograph probability distribution.

Although average flows are of interest, floods and low flows are primary causes of concern because of the threat to life and property of floods and threats to ecosystems and navigation of low flows. Whereas individual flows tend toward a normal distribution, extreme flows are best described by distributions such as the log-normal, Gumbel, and Pearson Type III. Figure 8-6 shows several peak flow distributions as computed by PeakFQ, a computer program distributed by the USGS. Such distributions are used to estimate probability and recurrence intervals for various flood levels. For example, from Figure 8-6, note the pick off of the 1% exceedance probable discharge is about 26,000 cfs, and the 95% exceedance probable discharge is about 1,800 cfs.

Floods are usually labeled in terms of their recurrence interval—the average time between floods of that magnitude or greater. Thus, a 1% probable flood is called the 100-year flood because it has a one-in-100 chance of occurring in any given year. Unfortunately, many people confusedly interpret the term to mean that such floods will occur every 100 years, which is dangerously wrong. The 1% probable flood can occur more than once a year or not happen again for 200 years. It's only over thousands of years that the one in 100 actually average out.

Low flows typical measures include the base flow and statistical versions like the 7Q10. Base flow is the minimum stream flow to which the stream returns between freshets, and in smaller streams it is governed by groundwater inflow to

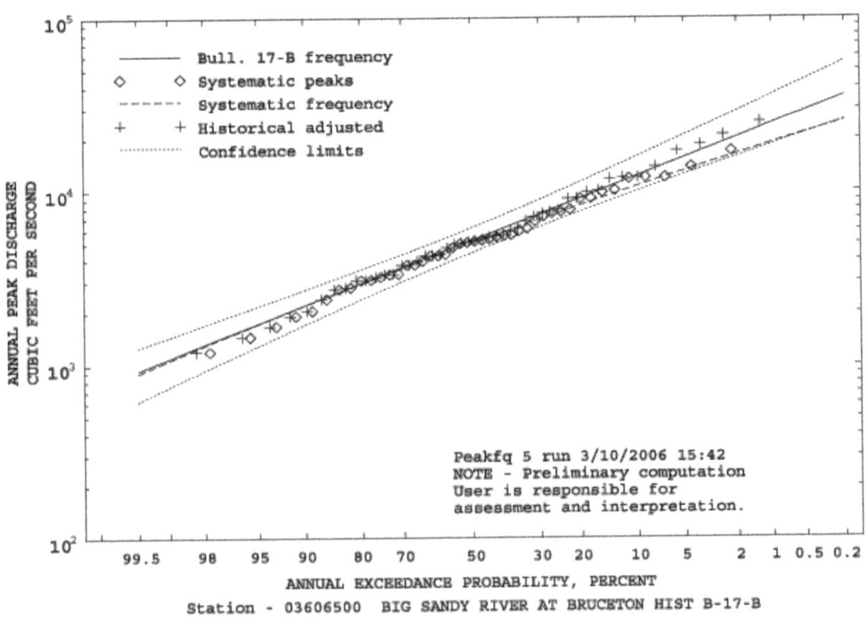

Figure 8-6. Example peak flow statistical distributions.
Source: Courtesy USGS.

the stream. The 7Q10, the lowest average discharge over a seven-day period occurring on average every 10 years, is commonly used in environmental evaluations, but other periods, such as the 30Q2, are sometimes preferred. More recently, flow requirements have been defined in more general terms, such as that suggested by the Brisbane Declaration: "Environmental flows describe the quantity, timing, and quality of water flows required to sustain freshwater and estuarine ecosystems and the human livelihoods and well-being that depend on these ecosystems" (International Water Centre 2016).

8.4.4 Water Control

Management of water levels and flows, termed *water control*, consists of monitoring and regulating waterways for their intended purposes, including navigation.

8.4.5 Further Information

Additional information on hydraulics can be found in publications such as Julian (2002) and French (1985).

8.5 SEDIMENTATION

Sedimentation processes—the erosion, transport, deposition, and consolidation of sediment—driven by wind and flowing water cause significant issues in inland waterways. Excessive erosion erodes land and streambeds, undercuts structures, and increases turbidity. Excessive deposition can restrict navigation and increase flood levels. Like water, both too much sediment and too little sediment introduce problems, such as navigation restrictions and/or and ecological habitat degradation. The designer's ideal is a stable, live-bed channel—one in which sediment moves through without substantial deposition or erosion.

8.5.1 Sediment Types

Sediment consists of inorganic mineral grains plus organic material. Figure 8-7 illustrates some typical sediment types. The inorganic component is typically classified by effective diameter of the grains, as depicted in Table 8-1.

Organic materials include of plant and animal fragments, algae, bacteria, and excreta. Bacterial polysaccharide excretions are particularly significant in fine sediment behaviors as they increase aggregation into larger composite particles. Algal mats on channel bottoms can effectively armor them against erosion by all but the strongest flows.

Cohesion dominates the behavior of the smallest sediment grains—clays and some silts—because single grains settle very slowly, if at all. Only when combined into larger aggregates of hundreds to millions of individual grains do they begin to settle at significant speeds. Cohesion, caused by electrophysical forces, bonds individual grains in aggregates through flocculation processes. Table 8-2 provides the increasing effect of cohesion with decrease in grain size.

Figure 8-7. Photographs of sediment types. From top: organic mixture, sand, gravel.

8.5.2 Transport

Sediment transport is often classified according to the three categories depicted in Figure 8-8, by the mode of transport, presence in the bed, and measurability. Each category includes two divisions, and the sum of those two divisions is always equal to the total sediment load, or transport rate, in mass or volume per unit time. Note the differences in relative sizes of the divisions within each category.

Transport mode category includes suspended load, the portion of the total load that is rarely or never in contact with the bed and travels at about the same speed as the water, and bed load, the portion that travels by sliding or hopping along the bed. Because bed load material may spend some time resting on the bed before being mobilized and accelerated, its average speed will be less than the water speed, and perhaps much less.

Bed presence category consists of wash load and bed material load. Wash load consists of sediment that is not present in the bed to an appreciable degree, and bed material load is present in the bed. Despite the seeming similarity to the transport mode categories, note that the wash load in Figure 8-8 is smaller than the suspended load, as bed material can be transported as suspended load. Some practitioners make the mistake of ignoring wash load in channel shoaling problems, reasoning that if it is not in the bed, it cannot deposit. However, the

Table 8-1. Sediment Grain Size Classification.

Class Name	Size Range				Approximate Sieve Mesh Openings per Inch	
	Millimeters	Φ	*Microns*	*Inches*	*Tyler*	*US standard*
Very large boulders	4096 ~ 2048			160 ~ 80		
Large boulders	2048 ~ 1024			80 ~ 40		
Medium boulders	1024 ~ 512			40 ~ 20		
Small boulders	512 ~ 256	−9 ~ −8		20 ~ 10		
Large cobbles	256 ~ 128	−8 ~ −7		10 ~ 5		
Small cobbles	128 ~ 64	−7 ~ −6		5 ~ 2.5		
Very coarse gravel	64 ~ 32	−6 ~ −5		2.5 ~ 1.3		
Coarse gravel	32 ~ 16	−5 ~ −4		1.3 ~ 0.6	2 ~ 1/2	
Medium gravel	16 ~ 8	−4 ~ −3		0.6 ~ 0.3	5	5
Fine gravel	8 ~ 4	−3 ~ −2		0.3 ~ 0.16	9	10
Very fine gravel	4 ~ 2	−2 ~ −1		0.16 ~ 0.08	16	18
Very coarse sand	2.000 ~ 1.000	−1 ~ 0	2000 ~ 1000		32	35
Coarse sand	1.000 ~ 0.500	0 ~ 1	1000 ~ 500		60	60
Medium sand	0.500 ~ 0.250	1 ~ 2	500 ~ 250		115	120
Fine sand	0.250 ~ 0.125	2 ~ 3	250 ~ 125		250	230
Very fine sand	0.125 ~ 0.062	3 ~ 4	125 ~ 62			

(Continued)

Table 8-1. Sediment Grain Size Classification. (Continued)

Class Name	Size Range				Approximate Sieve Mesh Openings per Inch	
	Millimeters	Φ	Microns	Inches	Tyler	US standard
Coarse silt	0.062 ~ 0.031	4 ~ 5	62 ~ 31			
Medium silt	0.031 ~ 0.016	5 ~ 6	31 ~ 16			
Fine silt	0.016 ~ 0.008	6 ~ 7	16 ~ 8			
Very fine silt	0.008 ~ 0.004	7 ~ 8	8 ~ 4			
Coarse clay	0.004 ~ 0.002	8 ~ 9	4 ~ 2			
Medium clay	0.002 ~ 0.001		2 ~ 1			
Fine clay	0.001 ~ 0.0005		1 ~ 0.5			
Very fine clay	0.0005 ~ 0.00024		0.5 ~ 0.24			

Source: Garcia (2007).

Table 8-2. Cohesion for Various Size Classes.

Size range (μm)	*Classification*	*Degree of cohesion*
> 62	Coarse-grained	Cohesionless
62 to 40	Fine-grained: coarse silt	Practically cohesionless
40 to 20	Fine-grained: coarse silt	Cohesion increasingly important with decreasing size
20 to 2	Fine-grained: medium and fine silt	Cohesion important
< 2	Fine-grained: coarse, medium and fine clay	Cohesion very important

Source: Garcia (2007).

distinction between wash load changes from reach to reach of a waterway, so upstream wash load can become bed material load in a downstream reach.

Finally, Figure 8-8 illustrates that total load can be divided into measured load and unmeasured load, which are self-defining. The distinction is necessary to remind sediment data users that a measured sediment discharge doesn't provide the complete picture and the unmeasured load must be estimated to obtain the total load. Because unmeasured load usually includes all or part of the bed load,

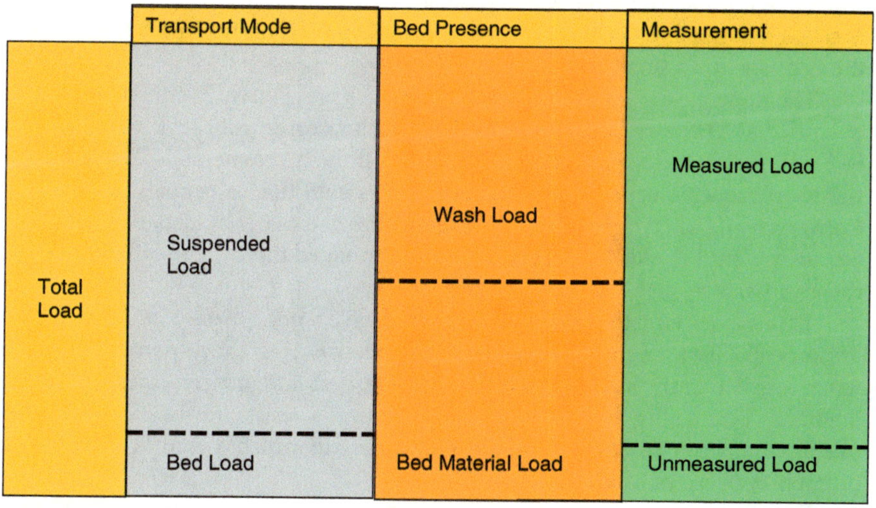

Figure 8-8. Sediment transport categorization.
Source: Adapted from USACE (1995).

they are sometimes confused. The relative size of the unmeasured load shrinks as measurement technology continues to improve.

A given flow (depth and velocity) has an inherent capacity to transport a certain bed material load of any given grain size. For example, a flow of 3 ft per sec at a depth of 1 ft can transport a substantial load of very fine sand but can transport no large cobbles. For a typical riverbed consisting of a range of sand sizes, there is a total transport capacity that is the sum of the transport capacities of every grain size. That transport capacity is the same whether the bed contains sediment or not; that is, the transport capacity is the same for each grain size even if the waterway is a concrete channel with no sediment on the bed.

Wash load has no comparable transport capacity. Traveling only as suspended load until flow conditions allow it to become bed material load, the amount in transport is the product of flow speed times sediment concentration at every point in the flow cross section. Any concentration up to a limit of several thousand kilograms per cubic meter can be transported as wash load/suspended load, with the limit reached when the concentration alters the nature of the flow. That condition is rare in navigable waterways.

8.5.3 Erosion, Deposition, Consolidation

The cycle of erosion, transport, deposition, and consolidation may be repeated many times for a given sediment as it moves downstream. For noncohesive bed load, the cycle repeats within minutes or seconds as grains are lifted into the flow, settle, and return to the bed, perhaps only inches or feet (centimeters or meters) from the liftoff spot. If fluid forces (lift and drag) remove significantly more sand grains than return by settling, net erosion of the bed surface occurs. If significantly fewer grains are removed than are returned, net deposition occurs. The case in which the number of grains leaving the bed equals the number settling to the bed is called a live stable bed, or dynamically stable bed. If no transport is occurring, the bed is simply stable.

The noncohesive erosion/deposition process previously described can also be described as a balance between transport capacity, aforementioned, and the actual bed material load at a waterway cross section. If the bed material load exceeds the transport capacity, deposition to the bed occurs until the load equals the capacity. Conversely, if the transport capacity exceeds the bed material load, erosion of the bed occurs until the load equals the capacity, provided there is enough sediment of the given size available at the bed surface.

This capacity of the flow to transport a given noncohesive grain size, with greater capacity for smaller sizes, produces the armoring effect. Armoring occurs when smaller grain sizes are lifted from the bed and not replaced, leaving a gradually coarsening bed surface that ultimately resists further erosion. The phenomenon can be seen in stream gravel bars in which a thin layer of gravel or cobbles overlays a bed of sand.

Fine, cohesive sediments can be eroded from the bed and transported great distances as suspended load and wash load before depositing. Figure 8-9 illustrates the cycle. Erosion of bed material occurs primarily when the stresses exerted by the

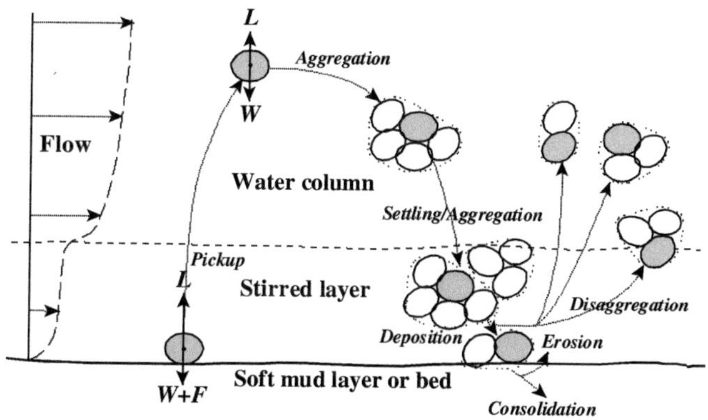

Figure 8-9. Cohesive sediment transport cycle in profile.

flow on the bed exceed a critical value for erosion. Unlike noncohesive sediments, cohesive beds do not experience significant simultaneous erosion and deposition. Instead, the bed is stable, with no significant erosion or deposition; erosional, without substantial deposition occurring; or depositional, without substantial erosion occurring.

Erosion of cohesive beds may occur as a few particles picked off at a time, called *surface erosion*, or by removal of a layer of bed, called *mass erosion*. A third mechanism, fluidization with entrainment, is discussed following. Surface erosion and mass erosion occur when the combined flow stresses on the bed exceed a critical value that is characteristic of the bed material. The critical shear stress for surface erosion is smaller than that for mass erosion, and both are smaller than the nominal shear strength of the bed as defined by traditional soil mechanics tests (penetrometer, triaxial shear). Values for erosion critical shear stresses can be determined by flume and field experiments or by rheological testing.

Settling and deposition of cohesive sediment is dominated by flocculation processes—aggregation of individual mineral grains into flocs. Bound together by physicochemical forces, flocs either grow larger or break apart through collisions with other flocs, producing a spectrum of sizes. They may be as small as a few microns, barely larger than the mineral grains themselves, or as large as a sand grain, albeit with a much lower density. Floc densities are much lower than the mineral density of the grains, as they include light organic material and substantial water in the spaces between grains.

If a floc grows large and heavy enough to settle through the turbulent flow field, it can approach the bed and become available for deposition. For a floc to deposit its internal shear strength—the stress as which the intrafloc bonds will break—must exceed the stresses applied by the high shear zone near the bed, by collisions with other flocs, and by collision with the bed. The rate of deposition is usually treated as a probabilistic process: If the shear stress exerted by the flow is above some threshold value, deposition does not occur, but below that threshold the rate of deposition is related to the settling velocity, concentration, and shear stress exerted on the bed by the flow.

Consolidation of deposited sediment occurs as pore water flows out of the bed and the grains pack together more closely. Sand-sized and larger sediment grains can consolidate to a small degree by rearrangement of the grains to fill voids, but silts and clays and sediment mixtures exhibit much larger voids, which can collapse, expelling water, increasing density, and decreasing volume substantially. Consolidation occurs over hours to decades and is accelerated by overburden pressure, such as new layers of sediment laid down on the bed surface and crushing the existing layers.

8.5.4 Bankline Erosion

Bankline and shoreline erosion are common environmental problems in waterways. Rivers naturally tend to meander, eroding one bank and depositing on the opposite bank in response to the twisting motions of turbulent flows. Human-induced changes to the landscape and waterway can also cause banking erosion by increasing or decreasing sediment yield to the river, by increasing or decreasing peak flow rates, and by changing the slope. Any of these alterations can lead to channel instabilities—bankline erosion and stream widening or deepening.

8.5.5 Further Information

Additional information on sedimentation can be found in publications, such as Vanoni (2006) and Garcia (2007).

References

Anderson, R. S., and S. P. Anderson. 2010. *Geomorphology: The mechanics and chemistry of landscapes*. Cambridge, UK: Cambridge Univ. Press.

Bedient, P. B., W. C. Huber, and B. E. Vieux. 2012. *Hydrology and floodplain analysis*. Upper Saddle River, NJ: Prentice Hall.

Clarke, R. T. 1998. *Stochastic processes for water scientists*. New York: Wiley.

French, R. 1985. *Open channel flow*. New York: McGraw-Hill.

Garcia, M. 2007. *Sedimentation engineering: Processes, measurements, modeling, and practice: MOP 110*. Reston, VA: ASCE.

International Water Centre. 2016. *Brisbane declaration*. Accessed June 8, 2016. https://www.watercentre.org/news/declaration.

IPCC (Intergovernmental Panel on Climate Change). 2014. *Climate change 2013—The physical science basis: Working group I contribution to the 5th assessment report of the intergovernmental panel on climate change*. Geneva: IPCC, UNEP.

Julian, P. Y. 2002. *River mechanics*. Cambridge, MA: Cambridge Univ. Press.

Poff, N. L., J. D. Allan, M. B. Bain, J. R. Karr, K. L. Prestegaard, B. D. Richter, et al. 1997. "The natural flow regime." *Bioscience* **47** (11): 769–784.

Ponce, V. 1994. *Engineering hydrology*. Upper Saddle River, NJ: Prentice Hall.

USACE. 1995. *Sedimentation investigations of rivers and reservoirs*. EM No. 1110-2-4000. Washington, DC: USACE.

Vanoni, V. A. 2006. *Sedimentation engineering*. MOP 54. Reston, VA: ASCE.

CHAPTER 9
Dredging and Disposal

9.1 GENERAL

The USACE annual dredging efforts for 2009 through 2014 average about 2.25 million cubic yards (yd^3) annually, costing about $1 billion per year. This dredging includes both coastal ship channels and harbors (saltwater) and inland ship and barge channels and harbors (brackish and freshwater). The COE annual pipeline contract dredging bids from 2009 to 2014 ranged from approximately 60 million to about 120 million yd^3, mechanical bucket dredging contracts ranged from approximately 10 to 20 million yd^3, and WID was bid once between the period of 2009 to 2014 for approximately 22,000 yd^3 in 2010 (USACE 2015b). More than 630 million tons of cargo is transported using barges within the approximate 25,000 mi of inland and intercostal waterways (USACE 2009b). If sediment deposition is impeding navigation and requires removal, the sediment should be used to support other missions, such as flood control or ecosystem restoration, when feasible (USACE 2015a). Considering dredged sediment (dredged material) as a resource, rather than a waste, is supported by the concept of sustainability.

Dredging quantities are much larger for the alluvial systems than the "clear-water" rivers. Dredging in the inland waterways is performed primarily with hydraulic pipeline equipment or mechanically by cable-suspended buckets. Cutterhead pipeline dredges are used extensively on the Intracoastal Waterway and wherever upland CDF are available, such as GIWW, Delaware River, James River, and Savannah River, particularly when open-water placement sites are not cost-effective. Urban harbors and channels are typically dredged mechanically due to equipment mobilization/demobilization; distance between the dredging location and CDF; access to the CDF; or because the CDF has limited capacity for water management, such as Lake Erie in New York, Los Angeles, Seattle, and San Francisco.

Dredging in marine waters (saltwater dredging) and ocean disposal is outside the scope of this manual. However, this subject is a candidate for inclusion in a future Manual of Practice, *Environmental Sustainability of Ship Channels.*

9.2 DREDGING, CONVEYANCE, AND PLACEMENT TECHNIQUES

For purposes of this technical report, *dredging* is defined as the action of excavation and relocation of sediments to clear a navigation channel, turning basin, ship berth, or marina. Dredging is performed using mechanical equipment, hydraulic equipment, or a combination of both. Dredging removes sediment from recent deposits in a navigational waterway (maintenance) or from a long-time natural state to create new navigation access (new-work construction). Excavated material will range from large rock to sand, silt, and soft mud (composed of silt and clay aggregated particles). Harder materials such as rock and compact glacial materials are more difficult to remove. Rocky materials, such as granite, sandstone, limestone, or coral, are typically removed only when new channel deepening or widening occurs. Other materials such as cobbles, gravel, sand, and mud are removed routinely during maintenance dredging. Coarser materials occur where the surface water hydrology is dynamic, as where fast currents or waves influence the sediment transport regime. Inland maintenance materials are usually relatively soft sediments available for removal.

The dredging project is a function of site conditions, dredged material type and quantity, equipment availability, operations, and logistics for the discharge of dredged sediment. The dredging project can be divided into four components: (1) dredging, (2) conveyance, (3) placement, and (4) dredged material management. Each of the four components is integrated into the overall dredging project. Dredging, conveyance, and placement techniques are addressed first in this chapter, then the dredged material management component is described in a separate section of this chapter.

Environmental dredging generally refers to contaminant remediation or cleanup projects where removal of contaminated sediment from the waterway to enhance environmental quality is the primary objective (Palermo et al. 2008). Although the United States has made great strides in reducing the quantity and type of wastes discharged to the aquatic environment, ongoing sources continue to enter aquatic environments, in addition to contaminants from historical activities. Overall, however, the trend in harbor sediment quality has clearly been improving as a result of the Clean Water Act (CWA), wastewater discharge programs, and other initiatives. The CWA is frequently used as a short form of the Federal Water Pollution Act of 1972 (see Chapter 5).

9.2.1 Dredging Equipment

Dredging is performed using mechanical equipment, hydraulic equipment, or a combination of both. The proper dredging equipment will be determined by consideration of

- Equipment availability,
- Dredging site conditions/site access,
- Material type,

- Placement site constraints,
- Distance to placement site, and
- Volume and type of material to be dredged.

Although not ideal, equipment availability can be the deciding factor for dredge equipment selection. For example, limited material barges may suggest a smaller dredging production rate, or a larger-than-preferred pipeline dredge may require a larger settling basin or regular work stoppages to allow for settling time. If limited equipment will be available during an in-water work window (described in Section 9.4), the design of the dredging project may require modifications. Therefore, knowledge of locally available equipment is important for the design of the dredging project.

Site conditions, such as limited access, wave energy, shipping or recreational craft traffic, or limited draft, must be considered, as well as dredged material type that could require specific equipment. For example, consolidated sediment may require a digging clamshell bucket rather than a level-cut closed-lip clamshell bucket.

Distance to the placement site may dictate dredging and conveyance equipment, whereas dredged material management requirements, if specified, may dictate placement techniques. For example, hydraulic conveyance is typically most practical for an upland CDF; therefore, hydraulic dredging should be considered as a first alternative. However, if the placement site is in-water several miles from the dredging location, barge transportation may be preferred, and therefore mechanical equipment should be considered first. Hydraulic conveyance is almost always achievable, but pump requirements and pipeline pathway may reduce feasibility. The volume of sediment to be dredged can influence the type of dredge for the job. For example, a dredging project of 20,000 yd^3 or less would not typically justify the mobilization and setup of a hydraulic dredge; therefore, a mechanical dredge will likely be the alternative of choice, unless more than one dredging project can be combined.

Mechanical

Mechanical dredges consist of clamshell buckets operated from a derrick and controlled by wire cables (Figure 9-1) and fixed arm excavators (standard and long-reach) mounted on a barge (Figure 9-2). Each type of mechanical dredge has several variations. For example, the clamshell bucket may be a digging bucket with teeth, a closed-lip clamshell bucket, or an environmental bucket capable of a level cut. The backhoe bucket is typically open at the top but also includes a closed-top bucket like the horizontal profiler bucket.

The bucket ladder dredge and the drag-line dredge are also included in mechanical dredges; however, these are used in the mining industry and typically not used for inland navigation or a small-harbor navigation dredging.

Hydraulic Dredging

Hydraulic dredges use a centrifugal pump that excavates the sediment and transports it as fluid slurry to a discharge area. The most common types of

Figure 9-1. Mechanical bucket dredge.
Source: USACE (2015a).

Figure 9-2. Backhoe dredge New York.
Source: Courtesy of Great Lakes Dredge and Dock Company, Oak Brook, Illinois (USACE 2015a).

hydraulic dredges are trailing suction hopper dredges and cutterhead pipeline dredges (Figure 9-3). Less common types are dustpan (plain suction) and side-caster (trailing suction) dredges. Special-purpose dredging systems have been developed during the last few years to pump dredged material slurry with high

Figure 9-3. Hydraulic cutterhead dredge.
Source: USACE (2015a).

solids content and/or to minimize the resuspension of sediments. For example, a mechanical dredge will place sediment in a barge, and the specialty hydraulic dredge will slurry the sediment for conveyance to the placement area. Although special-purpose systems are not commonly used on navigation dredging projects, they may be suited to dredging projects that involve contaminated sediment cleanup.

Agitation Dredging and Water Injection Dredging

As with each type of dredging, the purpose is to relocate sediment. Some dredging equipment may be used in a different manner to achieve similar results through different means. Agitation dredging and water injection relocate sediment using natural hydraulic energy, such as river currents. Agitation dredging involves two distinct phases: suspension of bottom sediments by some type of equipment and transport of the suspended material by currents. Agitation dredging methods include hopper dredge overflow, prop wash, vertical mixers or air bubblers, rakes or drag beams, and WID. Agitation dredging, sometimes called overflow dredging, is performed by a hopper dredge, allowing the sediment and water to flow over the hopper and redeposit in the estuary or river. Agitation dredging is different than typical hopper dredging in that the physical characteristic of the sediment is fine particles that will not settle and are released for natural transport in the water column to be accreted in areas such as subsiding wetlands or subtidal habitat. Drag beams, or bed levelers (Figure 9-4), are pulled along the bottom to mechanically loosen and relocate sediment from a high spot to either level the bed or suspend the sediment into the water column to be transported by natural currents, or a

Figure 9-4. Bed leveler suspended by an A-frame on a work barge.
Source: Courtesy of Bean Dredging Company, New Orleans (USACE 2015a).

combination of both (USACE 2015a). WID fluidizes the sediment by injecting pressurized water via water jets so the sediment will flow by gravity to a depth below the navigation depth (USACE 1993a). WID does not confine the sediment in any way but instead allows the sediment to flow under gravity into a deeper part of the channel or thalweg along the bottom of the channel due to its density being greater than water. Sidecast dredging is another method to use natural hydraulics to convey sediment rather than pumping or barging sediment downriver. Sidecast dredging is accomplished using a sidecast specialty dredge, or in some cases a hydraulic pipeline dredge, which discharges sediment into the river currents, also described as flow lane placement. In the cases of agitation dredging, sidecast, and flow lane placement, the sediment is confined only for a short time through the dredge pump and pipeline until the sediment is discharged into the water column for natural conveyance by hydraulic forces. In the case of WID, the sediment is not confined but instead allowed to be naturally carried by the river currents. In each of these types of dredging, nourishing the water column with fine sediment can be beneficial for regional sediment management. Some bays, such as the San Francisco Bay Estuary and Delta, have become "sediment starved," and maintaining sediment within the littoral system is a sustainable practice for regional sediment management.

9.2.2 Conveyance and Placement

Conveyance techniques are determined by the dredge type as well as the placement location and placement requirements. Mechanically dredged sediment is typically conveyed using material barges. However, makeup water may be added to mechanically dredged sediment to create a slurry that can be hydraulically conveyed through a pipeline to the placement site. If sediment is hydraulically dredged, then rarely is the material conveyed by a barge or truck unless it is first dewatered in a rehandling facility.

Placement location and conditions might determine the conveyance method and potentially the dredging method. For example, if dredged sediment is to be placed into an upland CDF, which contains ample capacity for settling and effluent discharge and is within approximately 1.6 km (1 mi) from the dredging site, hydraulic dredging and conveyance will be an initial consideration. Alternatively, if dredged sediment placement will occur several miles from the waterway to be dredged and the site does not have dewatering capabilities, then the material will likely require mechanical placement, and mechanical dredging will be an initial consideration. Placement location, with respect to the water surface (upland, nearshore, ocean placement), can also influence conveyance and dredging techniques. Placement requirements, such as wide dispersion in thin lifts for wetland nourishment, can determine dredge type. In the case of beneficial use for wetland nourishment, hydraulic placement will require the least amount of sediment handling. Therefore, hydraulic dredging will be initially considered. Long-distance hydraulic conveyance has been successfully performed with multiple booster pumps to reach long distances for wetland nourishment.

9.3 DREDGED MATERIAL MANAGEMENT

The sediment dredged from navigable waterways for the purpose of navigation is an important resource that is vital to a robust, resilient, and sustainable ecosystem. Although sediment is a resource, it is often considered a waste that requires disposal. Instead, when cost-effective, the sediment should be managed for a beneficial purpose. Because the millions of cubic yards of material removed during navigational dredging each year is nearly 100% sediment (sands, silts, clays, and water), dredged material can also be referred to as "dredged sediment." In-water placement should be the first consideration for cost-effectiveness and environmental sustainability. Removal of this valuable sediment resource from the aquatic system poses a permanent loss of benefit to the littoral system and therefore is not a sustainable practice; efforts should be made to maintain the sediment in the littoral system when cost-effective.

9.3.1 Long-Term Management

An LTMS, DMMP, or DMMS describes a plan for managing sediment removed during maintenance dredging.

For the USACE, ER 1105-2-100, *Planning Guidance Notebook* (USACE 2000), describes a process to develop a DMMP for a period of 20 years. During project development, a base plan for management of dredged material is established. The base plan must be consistent with the federal standard, which is defined as the dredged material management alternative or alternatives identified by the USACE that represents the least costly alternative consistent with sound engineering practices and meeting environmental standards established by the CWA 404(b)(1) evaluation process or ocean dumping criteria (33 CFR 335.7).

Three guidance documents should be referenced for dredged material management of inland navigation, which will be placed under the regulatory authority of the CWA:

1. The technical framework provides overall guidance for dredged material management: *Evaluating Environmental Effects of Dredged Material Management Alternatives—A Technical Framework* (EPA 842-B-92-008) (EPA/USACE 2004).
2. The inland testing manual, USEPA/USACE guidance document *Evaluation of Dredged Material Proposed for Discharge in Waters of the U.S.—Testing Manual* (EPA/USACE 1998), provides guidance with testing, evaluation, and assessment.
3. The UTM guidance is provided for dredged material placement with return water to the waters of the US, which is regulated by the CWA: *Evaluation of Dredged Material Proposed for Disposal at Island, Nearshore, or Upland Confined Disposal Facilities—Testing Manual* (USACE 2003).

In the event the dredging occurs for inland navigation but the dredged material is placed under the authority of the MPRSA, a fourth document should be followed: *Evaluation of Dredged Material Proposed for Ocean Disposal, Testing Manual* (EPA 503/8-91/001) (EPA/USACE 1991).

These guidance documents provide information regarding testing evaluation and assessment of dredged material.

9.3.2 Dredged Material Management Categories

Twelve categories describe dredged material management. These categories have the advantage of including information regarding the *location* of placement relative to water elevation (upland, nearshore, inland waters, and ocean waters), as well as information regarding the *intent* of discharge (disposal versus placement for beneficial use). The location as well as the intent of placement can affect regulatory requirements.

- **Placement for upland land development**: placed for upland elevation gain or land development, including commercial, agriculture, and parks and recreation; also includes side casting for dike or levee construction, and land development of a CDF.
- **Upland placement for ecological habitat**: placed for upland habitat, such as upland forest or bird habitat; also includes transitioning an upland CDF for habitat.
- **Upland placement for soil reuse**: placed for reuse, such as upland CDF with the intention of reusing a portion of the dredged sediment for beneficial

purpose (e.g., rehandling for landfill cover, mine reclamation, construction fill, etc.).

- **Upland disposal**: placed in an upland CDF for disposal with no intention of present or future beneficial purpose. This option includes rehandling for disposal into a landfill but does not include reclamation uses, such as daily landfill cover.
- **Beach or nearshore placement for shoreline protection or beach nourishment**: placement on or along the shoreline (coastal and inland), including feeder berms. This option includes sediment placed directly for beach nourishment, as well as nearshore placement with the intent for the majority of sediment to remain within the depth of closure or littoral zone.
- **Shallow water placement for wetland, marsh, or habitat**: dredged material placed below ordinary high water for wetland or marsh nourishment/creation or other habitat, such as bottomland hardwood, salt marsh, swamp, wooded wetland, scrub-shrub, and forested wetland. Benefits can also include storm and surge protection.
- **Unconfined in-water placement (river, lake, and estuary) for RSM**: dredged material placed into a river, lake, bay, or estuary, including flow lane placement, sidecasting, overflow, agitation dredging, and other unconfined open water placement. Includes aquatic placement for future rehandling or transfer (ATF). Benefits are RSM sustainability.
- **Confined in-water placement for beneficial purpose**: in-water placement confined to a defined footprint, such as a nearshore CDF or CAD that will support subaquatic vegetation, essential fish habitat, or other beneficial use.
- **Confined in-water disposal**: in-water disposal confined to a defined footprint, such as a nearshore CDF or CAD with no present or future intent of beneficial use.
- **Island placement for benefits**: dredged material placed for island creation or island nourishment, including levees. Also included in this option are other categories that apply specifically to island placement, such as upland habitat, beach nourishment, and nearshore habitat. Does not include disposal (see disposal categories).
- **Ocean placement or deep-water lake for beneficial use**: ocean placement of dredged material into designated MPRSA site, intended for beneficial purpose.
- **Ocean disposal or deep-water lake disposal**: placement of dredged material into designated MPRSA site or CWA site, intended for disposal.

The intent of beneficial use is integrated into nine categories. Six of these nine categories are intended to provide benefits in the aquatic environment, whereas the three remaining categories are located in the upland environment (i.e., above ordinary high water). Beneficial use of dredged sediment is achieved when used as a resource in a productive way which provides environmental, economic, or social benefits (USACE 1987a, Childs 2015).

9.3.3 Dredged Material Characterization and Management

Characterization of dredged sediment will occur prior to dredging and typically will occur prior to executing a dredging contract. Characterization will be dependent on preferred and potential DMMS and current knowledge and previous characterization events. Regulatory requirements will depend on the DMMS, which may depend on physical, chemical, and biological characteristics of the dredged material. Understanding physical characteristics, such as grain size distribution and settling/consolidation, will assist in dredging and dredged material management design. For example, the settling velocity of the dredged material can constrain the volumetric info into a CDF (USACE 1987b). Understanding chemical and biological characteristics will allow for proper management of the dredged sediment (EPA/USACE 2004). The characterization strategy will be specific to each dredging project. For example, dredged sediment that is likely to be placed into an upland CDF will not be unnecessarily characterized for aquatic placement, or if the dredged prism is expected to be chemically segregated (via footprint or layering), the characterization effort should reflect this professional knowledge.

Because ecological function, as well as environmental regulations, can be divided into three location-specific alternatives with respect to water level, characterization is further described by the following: upland (terrestrial), nearshore (terrestrial/aquatic interface), and in-water (including inland and ocean waters).

Upland Placement of Dredged Material

As dredged material is placed into a CDF, the effluent, volatilization, and potentially leachate will be of immediate potential concern and should be considered in the first phase of evaluations. However, after dewatering, runoff and plant and animal uptake become potential pathways of concern for the dredged materials exposed at the surface. Therefore, only sediment that is representative of the dredged material that will be at the upper layer(s) of the CDF should be evaluated for runoff and uptake. As necessary, engineering controls, such as placement of a clean cover, can be considered prior to closure of a CDF.

The *Evaluation of Dredged Material Proposed for Disposal at Island, Nearshore, or Upland Confined Disposal Facilities—Testing Manual* (USACE 2003) describes six potential migration pathways for upland placement of dredged material: (1) effluent, (2) leachate, (3) volatilization, (4) runoff, (5) plant uptake, and (6) animal uptake.

Nearshore Placement of Dredged Material

Regulations associated with both the MPRSA (40 CFR Part 227 and Part 228) and the CWA (40 CFR Part 230) recognizes the "intent" for discharge of dredged material during a navigation dredging project. Dredged material can be discharged with the intent for "disposal" or discharged with the intent for "fill" (beneficial use). The regulatory authority within the territorial sea is described in the Ocean Testing Manual (OTM) (EPA/USACE 1998) and the Inland Testing Manual

(ITM) (EPA/USACE 2003). If the dredged material is placed inland, then 40 CFR Part 230 and the ITM will provided appropriate guidance. Intertidal and near-shore placement of dredged sediment is a unique opportunity to create habitat, nourish ecological habitat in subsiding marsh and mudflats, and provide storm protection.

In-Water Placement (Inland Waters or Ocean Waters)
Dredged material place in-water will be either regulated under the MPRSA with guidance in the OTM (EPA/USACE 1998) or under the CWA with guidance in ITM (EPA/USACE 2003), depending on the location of placement. If dredging occurs inland but disposal occurs in ocean waters, then the MPRSA is the regulating authority; if placement occurs inland, then the CWA is the regulating authority.

Potential positive and negative impacts should be evaluated. In-water beneficial uses of dredged sediment include enhancing the aquatic fish habitat, isolating existing contaminant in bottom sediments, or simply maintaining the sediment in the aquatic system for benefits. Potential negative impacts can occur when the placement site is near aquatic vegetation or shellfish habitat.

9.4 IMPACTS AND BENEFITS

Although management of typical navigational dredged material allows for few design specifications, these specifications can be critical in the managing impacts and engineering for benefits, such as ecological function or flood protection. The design specifications that can be potentially controlled during dredged material placement are

- Location of sediment placement, to the extent of cost-effective feasibility;
- Elevation of sediment (pre- and post-placement), including topographic/bathymetric variability;
- Thickness of newly placed sediment;
- Sequencing sediment type (physical and chemical characteristics); and
- Timing of placement.

For some projects, ongoing placement events will allow for ongoing modifications, or adjustments, to the design for improvement to ecological function. Adaptive management practices can be applied to optimize ecological function with postplacement activities, which might include planting, grading, or adding features such as woody debris or hydraulic control structures, depending on the design criteria of a project.

Protection against potential biological impacts occurs with environmental windows, which have been used as dredging project management practices since the early 1970s, following passage of the NEPA in 1969 (Suedel et al. 2008).

Environmental windows topically indicate times when in-water work *is not* allowed, whereas work window indicate times when in-water work *is* allowed. Discussions with the local regulators about risk-informed decision making should occur to appropriately manage the sediment.

9.5 RESUSPENSION

Resuspension is a common concern by regulators; however, little is truly known about potential effects of suspended sediment during dredging. Site-specific and project-specific conditions such as timing and duration of the proposed in-water work; presence, life stage, and locality of aquatic species and their habitats; and predicted impacts to the local ecology can be used in conjunction with state water quality regulations to allow for the development of a WQMMP. The WQMMP can include an effects-based turbidity threshold, designed to be protective of site-specific aquatic species and life stage, while allowing for the project-specific in-water work and meeting risk-informed regulatory requirements. Environmental windows may be established to eliminate the potential for biological impacts during dredging and placement activities.

Effects-based turbidity thresholds that are based on project-specific information and a dose-response model are typically less stringent than the statewide standards. Using a turbidity and total suspended solids (TSS) relationship, effects on local ecology can be predicted by a turbidity standard that is easily measured in the field. These predictive models can be updated as needed to include the project-specific information. The resulting project-specific turbidity thresholds, used in conjunction with prudent best management practices, can protect sensitive aquatic resources, satisfy regulatory statutes, and allow for an effective and efficient project.

Turbidity is often used by regulatory agencies as its surrogate for TSS. Turbidity is a measurement of light attenuation, which is affected by color as well as TSS, whereas TSS is a measurement of suspended particles. Turbidity is usually measured by *nephelometry*, the relative measurement of light scattered by a water sample 90 degrees to the incident beam (Lewis 1996). A linear correlation between turbidity and TSS can be expected when the physical properties of the suspended particles are constant (Gippel 1995), but they are seldom constant. The correlation of TSS and turbidity is dependent on site-specific factors, including particle size, shape, and color, plus materials dissolved in water, such as tannin. For this reason, it is necessary to obtain or create a specific turbidity/TSS rating equation for the project by season. If data are not available to develop a turbidity/TSS correlation prior to the start of the construction, significant data collection will be required. However, being able to predict TSS levels with real-time turbidity measurements will be valuable during construction.

To develop an appropriate turbidity threshold for any given in-water work activity, there are four primary needs for information: (1) the aquatic species of interest, (2) an estimate of maximum turbidity or TSS conditions that are anticipated, (3) an estimate of potential exposure duration, and (4) potential contaminants and concentration levels. In addition to these information needs, it is also important to work with the appropriate regulatory agencies to determine an acceptable level of protection and an acceptable level of risk.

9.6 REGIONAL SEDIMENT MANAGEMENT

Regulatory language described in 40 CFR 230 Subpart H, titled "Actions to Minimize Adverse Effects," states, "Using planning and construction practices to institute habitat development and restoration to produce a new or modified environmental state of higher ecological value by displacement of some or all of the existing environmental characteristics" [40 CFR 230.75 (d)].

This regulation invites an approach of integrated water resources management within a watershed or ecosystem. Ecological metrics will enable project managers and resource managers to compare two or more dredged material management alternatives, as well as assist regulatory decisions when dredged material management has potential for long-term ecological gain, although short-term impacts may occur from the discharge of dredged material.

9.7 HABITAT DEVELOPMENT USING DREDGED MATERIAL

The development of habitat is a potential beneficial use for dredged material. Habitat creation is the establishment of relatively permanent and biological productive plant and animal habitat. The four types of habitat that are potential candidates for dredged placement sites are marsh, upland, island, and aquatic.

Appendix C, *Habitat Development Using Dredged Material*, presents a description of this environmental enhancement of habitats.

References

Childs, J. L. 2015. "Dredged material management categories for tracking beneficial use." *Tech. Note ERDCTN-DOER-R22*. Vicksburg, MS: Engineering Research and Development Center.

EPA/USACE. 1998. *Evaluation of dredged material proposed for discharge in waters of the US—Testing manual*. EPA-823-B-98-004. Washington, DC: EPA/USACE.

EPA/USACE. 2003. *Evaluation of dredged material proposed for disposal at island, nearshore, or upland confined disposal facilities—Testing manual*. ERDC/EL TR-03-1. Washington, DC: EPA/USACE.

EPA/USACE. 2004. *Evaluating environmental effects of dredged material management alternatives—A technical framework*. EPA 842-B-92-008. Washington, DC: EPA/USACE.

Gippel, C. J. 1995. "Potential of turbidity monitoring for measuring the transport of suspended solids in streams." *Hydrol. Processes* **9** (1): 83–97.

Lewis, J. 1996. "Turbidity-controlled suspended sediment sampling for runoff-event load simulation." *Water Resour. Res.* **32** (7): 2299–2310.

Palermo, M. R., P. R. Schroeder, T. J. Estes, and N. R. Francingues. 2008. *Technical guidelines for environmental dredging of contaminated sediments*. Technical Rep. No. ERDC/EL TR-08-29. Vicksburg, MS: Engineer Research and Development Center.

Suedel, B. C., J. Kim, D. G. Clarke, and I. Linkov. 2008. "A risk-informed decision framework for setting environmental windows for dredging projects." *Sci. Total Environ.* **403** (1–3): 1–11.

USACE. 1983. "Dredging and dredged material disposal." EM 1110-2-5025. Washington, DC: USACE.

USACE. 1987a. "Beneficial uses of dredged material." EM 1110-2-5026. Washington, DC: USACE.

USACE. 1987b. "Confined disposal of dredged material." EM 1110-2-5027. Washington, DC: USACE.

USACE. 1993a. "Water injection dredging demonstration on the Upper Mississippi River." *Tech. Note DRP-3-10*. Vicksburg, MS: US Army Engineer Waterways Experiment Station.

USACE. 1993b. *Water resources development in Illinois*. Washington, DC: USACE.

USACE. 2000. "Planning guidance notebook." EM 1105-2-100. Washington, DC: USACE.

USACE. 2003. "Evaluation of dredged material proposed for disposal at island, nearshore, or upland confined disposal facilities—Testing manual." *Tech. Rep. ERDC/EL TR-03-1*. Vicksburg, MS: US Army Engineer Research and Development Center.

USACE. 2009a. *Implementation guidelines for section 2039 of the water resources development act of 2007 (WRDA 2007)—Monitoring ecosystem restoration*. Accessed July 9, 2016. http://planning.usace.army.mil/toolbox/library/WRDA/wrda07Sec2039a.pdf.

USACE. 2009b. *Inland waterway navigation, value to the nation*. Alexandria, VA: Institute of Water Resources. Accessed December 11, 2018. https://www.usace.army/mil/Missions/Civil-Works/Project-Planning/Legislative-Links/wrda_2007_impguide/.

USACE. 2015a. *Dredging and dredged material management*. EM 1110-2-5025. Accessed February 2, 2016. https://www.publications.usace.army.mil/Portals/76/Publications/EngineerManuals/EM_1110-2-5025.pdf.

USACE. 2015b. *Dredging information system (DIS)*. Alexandria, VA: Institute of Water Resources, Navigation Data Center. Accessed February 2, 2016. https://www.navigationdatacenter.us/dredge/drgtype.htm.

CHAPTER 10
Training Works

10.1 GENERAL

Many natural rivers are shallow and/or meandering with secondary side channels, backwaters, and seasonal flooded areas. As rivers meander and the separation between bendways increases, bars may develop in the intermediate crossing. When the spacing between bendways becomes large, alternate bars may develop, or a middle bar may be created in the downstream bendway and establish a secondary channel (see Figure 10-1).

To modify these rivers for navigation to provide adequate and dependable water depth for vessel passage, it is usually necessary to concentrate flow to one main channel as much as possible. To provide such water depths, the modifications usually include training structures such as revetments to limit the amount of active meandering that can take place in the future. Revetments structures are typically constructed of stone or in some cases with wooden piles to help maintain the desired channel alignment and eliminate channel meandering. In conjunction with the revetments, some type of contraction training structure such as spur dikes, groins, L-head dikes, bendway weirs, or any other type of contraction work may be required. For environmental sustainability of these modified rivers, the construction of revetments and dikes should maintain or increase aquatic diversity while preserving the integrity of off-channel habitats. The design and maintenance of dikes should be directed to reducing sediment accumulation to prolong the life of aquatic habitats and maintaining connections of abandoned channels to the river. Pokrefke et al. (2012) describes most of the various types of revetments and contraction works such as dikes used in the United States. Figure 10-2 illustrates a typical channel training layout.

The training structures would cut off seasonal water flow into backwaters and old secondary channels, which is typically the habitat of many aquatic species. Because dikes and groins were typically connected to the riverbank, these structures disrupt the movement of fish along the bank where lower river velocities occur. By modifying natural channels with these types of training structures and creating in many cases one stable channel, the areas available to various species of aquatic plants and animals becomes somewhat limited.

Therefore, the goal of an environmental sustainable waterway is to reconnect the separated backwater areas and side channels to the main channel. This can be

*

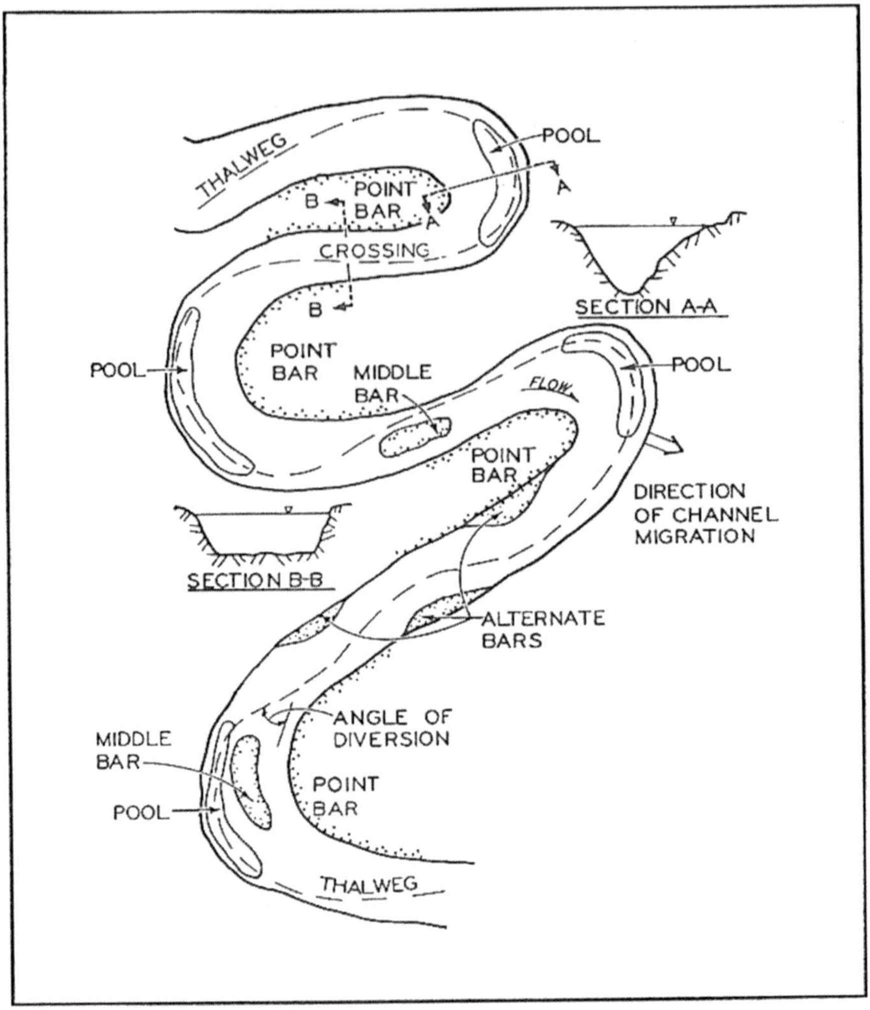

Figure 10-1. Channel planform.
Source: Pokrefke et al. (2014).

done by having low sections in levees or breaching the levee in selected locations. Riverbanks can be made more fish friendly by putting notches in dikes or leaving a gap in the dike (rootless dikes) to allow fish movement along the riverbank. The key to obtaining an environmentally sustainable waterway and integrating such modifications into the navigation channel is to ensure that the navigability of the waterway is also maintained. Often minor modifications can be made that do not reduce the efficiency or dependability of the navigation channel and do enhance the environment.

In these cases, the river engineer and fish and wildlife expert work together for the engineer to achieve enough channel contraction for sustainable navigation,

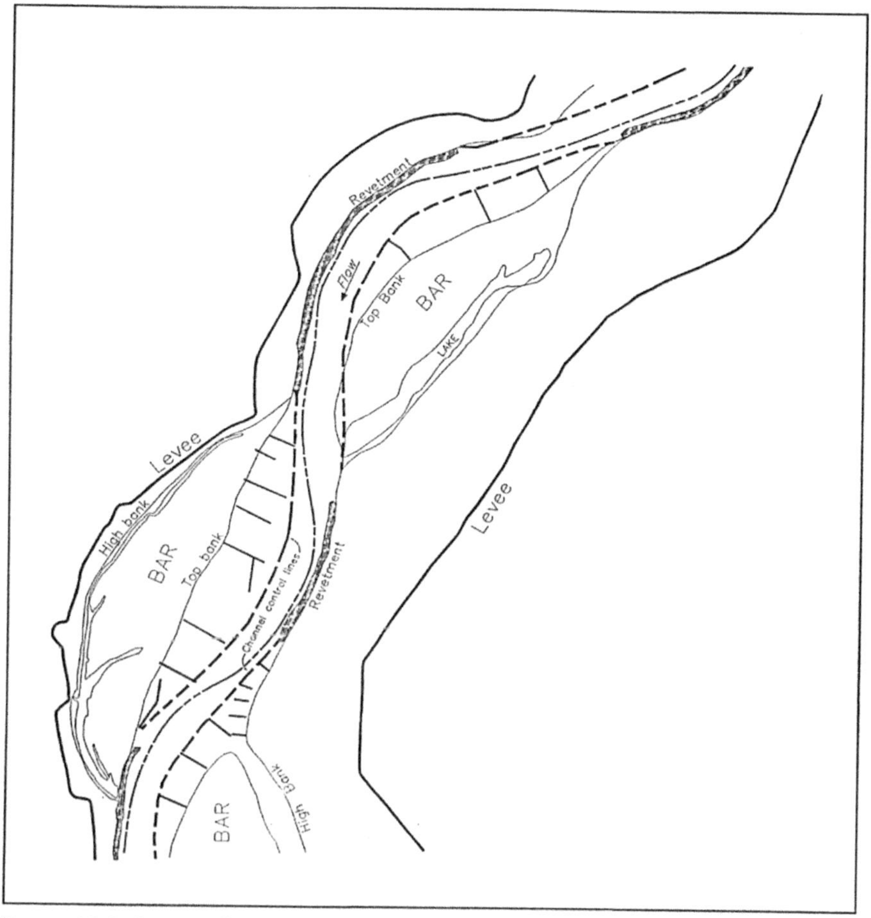

Figure 10-2. Layout for channel training structures.
Source: Pokrefke et al. (2014).

and the fish and wildlife expert gets the desired environmental enhancements by modifying the channel contraction works. In the US Army Engineer District, Omaha, more than 1,300 notches have been constructed either by removing stone or by omitting the repair of small portions of damaged river training structures along the Missouri River. The US Army Engineer District, St. Louis, has addressed the notching of Mississippi River dikes to create and support diverse habitats. For diversity, the designs work toward four primary river habitats: (1) fast water areas in the main river channel where the water current moves relatively quickly; (2) slow or quiet water areas that are outside of the main navigation channel; (3) wetted edge is constantly getting wet and later dry relative to river stages, constantly changing, and is important because there is a constant exchange of nutrients between land and aquatic environment; and (4) terrestrial or land separate from the shore (bars and islands), which provide protection from man and predators to the inhabitants.

10.2 ENVIRONMENTAL DESIGN MODIFICATIONS FOR TRAINING STRUCTURES

Examples of environmentally friendly designs are found in ASCE MOP 124, *Inland Navigation Channel Training Works* (see Pokrefke et al. 2012). In conjunction with those designs, the US Army Engineer District, St. Louis, been involved in a study titled *Environmental River Engineering Project on the Mississippi*, which also provides some of the examples that follow (USACE 2008).

10.2.1 Notched Dikes

> This type of dike has a section removed from the crown, primarily to alter flow and deposition/scour patterns for environmental enhancement. The notch is typically a minimum of 50 feet in width and 5 feet in depth. The location of the notch varies depending on the desired enhancement being developed. This is necessary to minimize the impacts on the river bank. Many times, notches are placed in multiple dikes and in a line to create a secondary channel for environmental enhancement. (Pokrefke et al. 2012, p. 117)

DuBowy (2012) reported that notches in dikes on the Mississippi River "foster diverse aquatic ecosystems. Various substrates, water depths, current velocities, and channel bottom configurations at fluctuating river stages all contribute to a diverse fishery community. This mosaic of micro-habitats results in a widely varied benthic invertebrate assemblage, and the large quantities of invertebrates then serve as food for fish populations throughout the system," p. 68.

Figure 10-3 is a schematic of a notched spur dike with a submerged extension. Figure 10-4 is a picture of three notched dikes from the Environmental River Engineering Project on the Mississippi River.

Pennington et al. (1988) report that more than 1,300 notches on the Missouri River had been installed or were created by not performing noncritical repairs. On certain reaches of the Mississippi River, the US Army Engineer District, St. Louis, has installed notches in entire stone spur dike fields. As those fields have responded to the notches and various flow conditions, the areas between the dikes have indeed developed a widely varied and desired topography that includes deeper pools immediately downstream of notches, slack water areas, increased wetted edges around sandbars, and terrestrial areas for birds and small animals. Figure 10-5 illustrates one of these dike fields and provides identification of the four primary river habitats discussed previously.

The location of a notch has an impact on the configuration of the bars and pools that form downstream of the notch. Shields and Palermo (1982) indicate that on the Missouri River, where the notch is located relative to the stream end of the dike influences the downstream bar location. In addition, the shape of the bankline will also influence the bar. When other types of dikes, such as L-head or longitudinal dikes, are modified with notches, the performance of those notches will also impact the bar formation and location (for information on these and

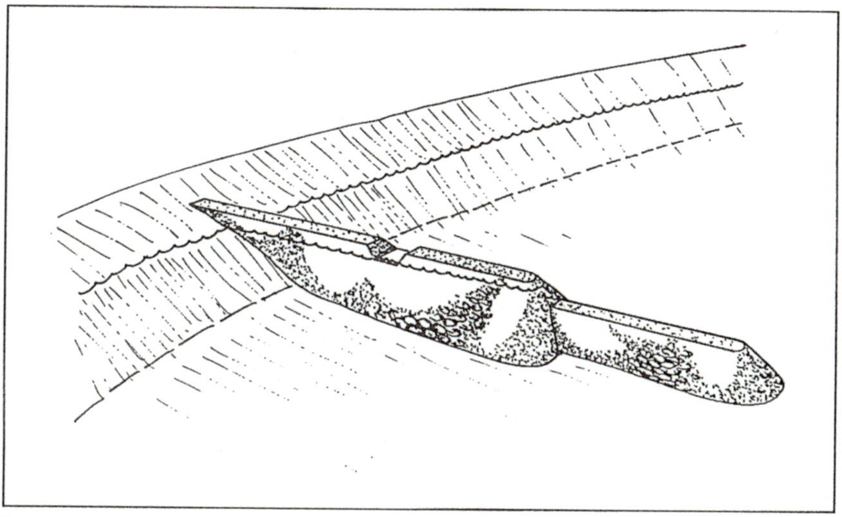

Figure 10-3. Stone spur dike with notch.
Source: Pokrefke et al. (2014).

Figure 10-4. Stone spur dikes with notches.
Source: USACE (2012).

other types of dikes, see Pokrefke et al. 2014). Figure 10-6 illustrates the differences that can take place. Therefore, if the river engineer or wildlife expert desires to design or modify a dike system, the trends shown in Figure 10-6 would be helpful to review.

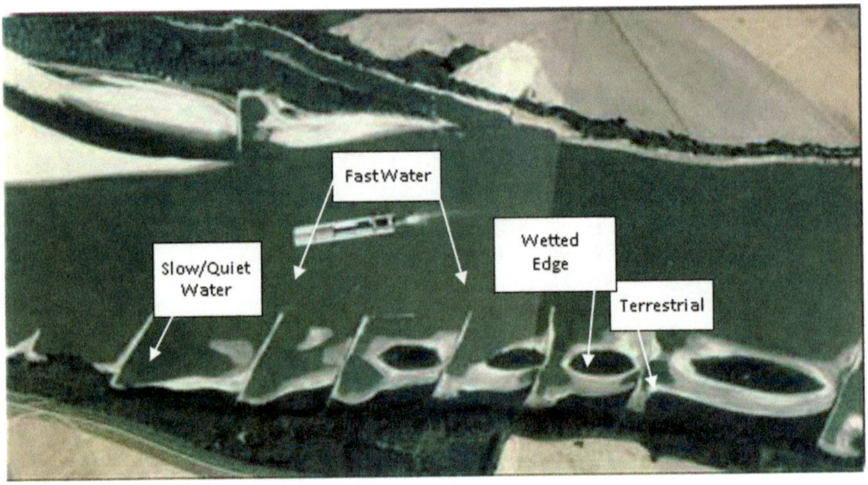

Figure 10-5. Stone spur dikes with notches.
Source: USACE (2008).

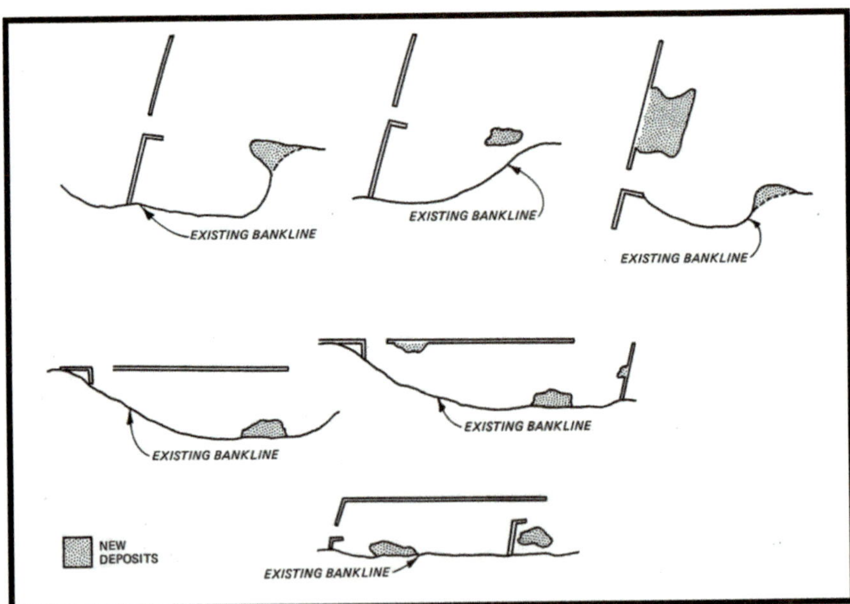

Figure 10-6. Sediment deposition patterns for dike with notches.
Source: Shields and Palermo (1982).

10.2.2 Rootless Dikes

The major difference between rootless dikes and classical spur dikes, groins, L-head dikes, or bendway weirs is that rootless dikes are not attached or keyed into the adjacent bankline. Typically, blunt-nosed chevron dikes and spur dikes not attached to the bankline (rootless spur dikes) fall into this category. The description for blunt-nosed chevron dikes is provided by the US Army Engineer District, St. Louis, and is presented in Pokrefke et al. (2012).

10.2.2.1 Blunt-Nosed Chevron Dikes

Blunt-nosed chevrons have an ellipse-shaped head and two legs that extend downstream, parallel to the flow, and are used to maintain the navigation channel. There are many cases in which chevrons have been used for navigation purposes, such as where split flows and channels are necessary in harbor or fleeting areas.

The size of these structures depends on the size of the river and location, the contraction width, and the purpose. Most chevrons on the Upper Mississippi River have footprints between 300 × 300 ft and 200 × 200 ft, which means legs 200 to 300 ft long and a width (leg to leg) of 200 to 300 ft (60.96 m × 91.44 m). The chevron has a top elevation set to allow overtopping during high flows (floods). The overtopping flows create a scour hole downstream from the structure, which provides excellent fish habitat. There are many variations of chevrons, typically for site-specific reasons and conditions or for additional environmental benefits. Variations include notches at the head of the structure, notches along one or both legs of the structure, differing leg lengths, parallel legs, and flared legs. Some chevrons are lined up parallel with the flow, whereas others are offset to one another. Figure 10-7 shows chevron structures in place on the Mississippi River.

Figure 10-7. Chevron dike field.
Source: Pokrefke et al. (2014).

10.2.2.2 Rootless Spur Dikes

"This type of dike has an offset from the river bank, meaning the structure starts some distance off of the bank. The typical offset distance is 100 ft or more. The river bank needs to be revetted a minimum of 100 ft upstream and 200 ft downstream if the bank is not bedrock. A rock pad or sill, minimum 30 in. thick is also commonly used in the width of the rootless section, 100 ft upstream and 200 ft downstream as well. The rootless section provides environmental diversity by altering flow and sediment transportation. Many times, multiple dikes are left rootless and in a line to create a secondary channel for environmental enhancement" (Pokrefke et al. 2014). Figure 10-8 shows a rootless spur dike on the Mississippi River.

10.2.3 Bendway Weirs

One type of river training structure that has been used by the USACE for about the last 20 years is bendway weirs. These structures were developed using a physical movable bed at the COE's Waterways Experiment Station in Vicksburg, Mississippi (see Derrick et al. 1994). Bendway weirs are submerged sills that are angled upstream and constructed in the navigation channel. Therefore, the towboat vessels navigate over the weirs as they move downstream. The bendway weirs use the energy in the river to direct flow away from the revetted bankline toward the point bar on the opposite side of the river. This helps erode the point bar somewhat, but more important it stabilizes the point bar to provide habitat for various species of birds and fowl. Pokrefke et al. (2012) provides both an in-depth explanation of bendway weir design parameters and a website for design.

As discussed in the model study report and documented later (see USACE 2012), very positive navigation and environmental benefits have resulted from the use of bendway weirs. In USACE (2012), concerning bendway weirs on the Mississippi River, it was stated,

Figure 10-8. Rootless spur dike.
Source: US Army Engineer District, St. Louis.

> The bendway weirs have not only provided navigation benefits, but many significant environmental benefits as well. A wider and more smoothly aligned navigation channel has resulted so traditional above-water dikes will no longer be built on the sandbars. Nesting habitat for the Least Tern, an endangered bird species is thus left largely undisturbed. Bendway Weir fields have also proven to provide habitat for a number of fish species. These environmental reefs have created diversity in the river bed and flow patterns in areas that were once narrow, deep, and swift. Monitoring efforts have shown that the federally-endangered pallid sturgeon use the weir fields for their habitat. (pp. 7–31)

Because bendway weirs are submerged all of the time, no prototype pictures are available; however, Figure 10-9 is an artist's rendering of a bendway weir field looking downstream.

10.2.4 Hard Points in Side Channels

The USACE (2008) found that construction of hard points, which are actually short stone spur dikes in side channels, is useful in stabilizing the riverbanks in that area. These hard points are keyed into the bankline and extend slightly into the river flow. There is no significant accumulation of sediments around the hard points; in fact, the hard points improve the habitat by creating scour holes. These deep plunge pools create a habitat that is attractive to catfish in particular. Figure 10-10 illustrates an application of hard points on the Mississippi River.

Figure 10-9. Artist's conception of bendway weir field.
Source: Pokrefke et al. (2014).

Figure 10-10. Hard points in a side channel.
Source: Pokrefke et al. (2014).

10.2.5 Off-Bank Line Revetment

There may be an instance where the river bank is caving in a location other than on the outside of a bendway, where the erosive mechanics of the flow can be very significant. Locations such as shallow water areas where bankline stability may be desired to protect property of some infrastructure require the river engineer to consider somewhat nontraditional bank stabilization. One design alternative would be off-bankline revetment. With this alternative, stone is placed essentially parallel to the bankline out in the water and is not keyed into the bank. This revetment is significant enough to reduce bankline erosion and provides diverse habitats for environmental considerations. The USACE (2012) has also notched some of these types of revetments to allow fish to move out of the faster-moving main channel currents into the slow currents between the revetment and bankline. The slower currents provide fishing areas that are attractive to fisherman. See Figure 10-11 for a schematic of off-bankline revetment.

10.3 ENVIRONMENTAL DESIGN MODIFICATIONS FOR CHANNEL ALIGNMENT

As a river meanders, over time the alignment of the bendways often become too tight for the design vessels to navigate the channel safely and efficiently. When such a situation occurs, it often becomes necessary to "cut off" the old, tight bendway to improve navigation. There are also flood stage benefits associated with a cutoff, but that is not discussed herein. These channel realignments are discussed in depth in Pokrefke et al. (2014; see their Section 4.5). In past years, the old bendway was left to natural processes with the upstream entrance to the old bendway filling relatively quickly with sands and coarser-grained sediments. The downstream exit from the old bendway tended to fill much more slowly with fine-grained sediments like silts and clays. Eventually the entrance and exit would be filled to an elevation high enough to eliminate average flows from the new main channel to pass into the old

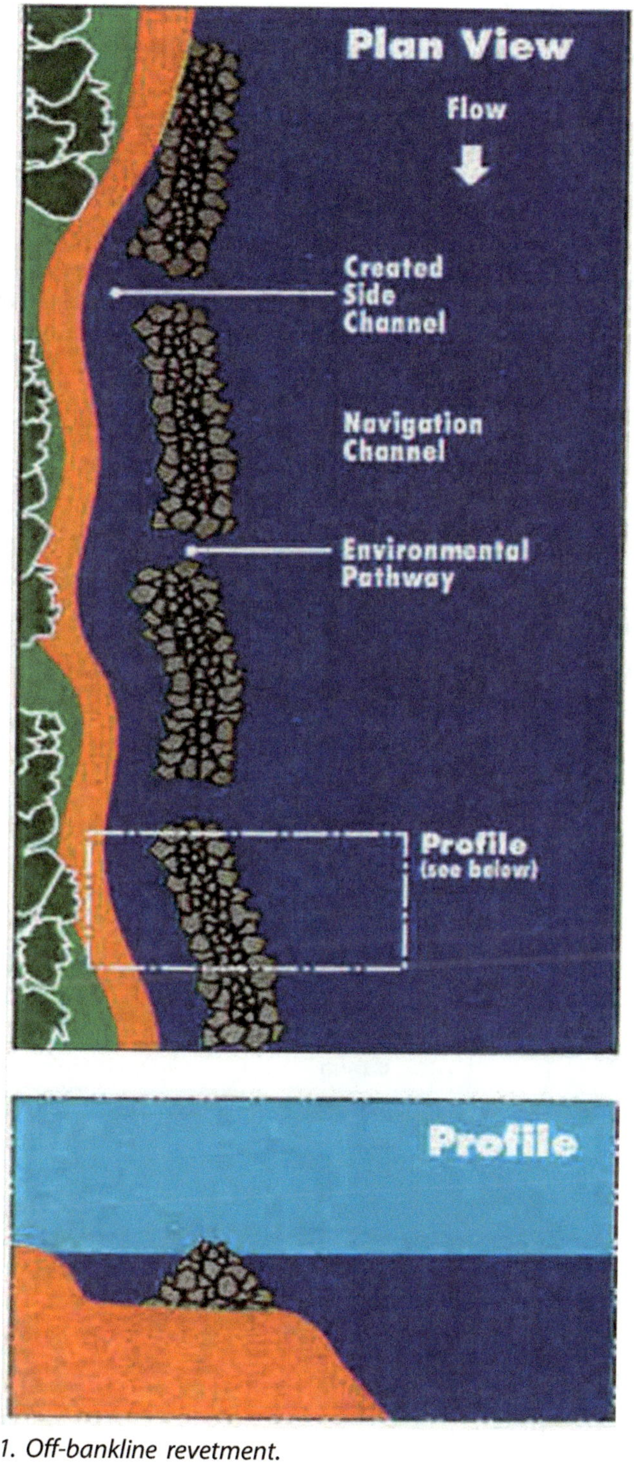

Figure 10-11. Off-bankline revetment.
Source: USACE (2012).

bendway. At this point, an oxbow lake would be formed, and quite often a levee or closure dike would be constructed across the entrance and exit, eliminating all flows from the main channel. Although such oxbow lakes were attractive to some, it was felt that when possible, maintaining a connection to the river helps to maintain a more diversified environmental situation.

Shields and Palermo (1982) state that the old bendway can be maintained with dikes to prevent sediment deposition. A closure dike or embankment in the upstream entrance and dikes at the downstream exit as shown in Figure 10-12 have worked well in numerous locations. The dikes at the exit narrow the channel to keep sediment moving. As river stages increase, flow backs into the old bendway at the downstream. As stages in the main channel decrease, the water accumulated in the old bendway will exit through the narrowed downstream leg and flush any sediment back into the main channel. Dike configurations such as this have proved to be very successful in creating backwater harbors and environmental areas that have remained open and useful for years.

10.4 ENVIRONMENTAL GUIDELINES AND DESIGN GUIDANCE

Pokrefke et al. (2012) provide an entire section on the issues and concerns that should be considered when doing an environmental enhancement to a navigation

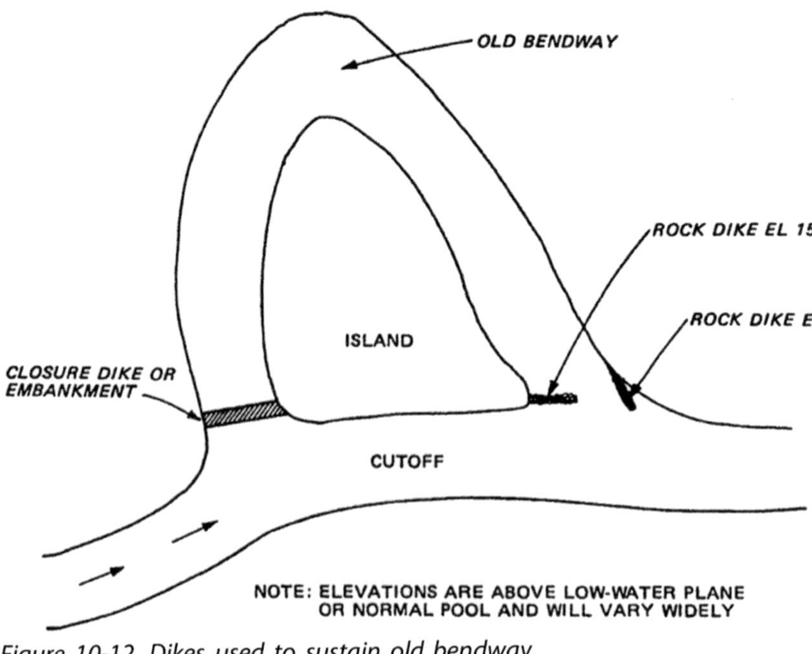

Figure 10-12. Dikes used to sustain old bendway.
Source: Shields and Palermo (1982).

project. The environmental guidelines and design guidance were developed by Burch et al. (1984) and are reproduced herein for the convenience of the reader.

> When providing environmental enhancements and modifications to the design of either new or existing river training structures certain aspects of the details to be included in the design need to be addressed. The designer should consider the general goals that are designed to be obtained by including environmental considerations and fold these goals into the design procedures for the river training structures. General environmental design goals and procedures will be addressed in this section.
>
> ***General goals***
>
> There are several environmental objectives and goals applicable to all dike design and construction:
>
> - Maintain or increase the aquatic habitat diversity by increasing the complexity of physical factors comprising the aquatic habitat.
> - Preserve the integrity of existing off-channel aquatic habitat area.
> - Schedule construction and maintenance to avoid peak spawning seasons for aquatic biota.
> - Design and maintain dike fields to prolong the lifetime of the aquatic habitat (e.g., reduce sediment accretion).
> - Maintain abandoned channels connections to the river.
>
> ***Design procedures***
>
> The ultimate configuration of the navigation channel and the locations of dikes and revetments to produce that channel are determined by the river master plan. . . . Master plan formulation to ensure incorporation of environmental considerations should include the following:
>
> - Formulate a draft river training master plan to achieve navigation, flood control, and bank erosion control objectives.
> - Using results of a habitat mapping study, evaluate the existing composition and spatial distribution of riverine habitats.
> - Using a multidisciplinary team, set general long-term goals for composition and spatial distribution of aquatic and terrestrial riverine habitats. These goals may be set for major reaches.
> - Modify the draft master plan to achieve these goals.
>
> A system of priorities based on anticipated results (in terms of habitat development) should be used to determine which structures should be modified first. The process as described for master plan formulation may result in recommendations to preserve and enhance dike field aquatic habitat. The following are suggested for design of a specific dike or dike field:
>
> - Evaluate the long-term potential of the dike field as aquatic habitat.

- Based on the above evaluation, determine whether design modifications or environmental features are in order.
- Consider manipulation of the basic dike design parameters to reduce the elevation of sediment deposition within the dike field.
- Qualitatively project the depths, velocities, and resulting substrates likely to occur in the dike field.
- Consider structural modifications to improve the aquatic habitat within the dike field.
- Consider management techniques to improve aquatic habitat within the dike field after construction.

Environmental features or modifications to dike designs have been applied on the Missouri River and the upper, middle, and lower Mississippi River. Dikes on the Missouri River contain the most environmental modifications and the techniques employed include notches, low-elevation dikes, vane dikes, and minimum maintenance practices. Notches are the most common, over 1,600 having been constructed. Environmental features occurring on the upper Mississippi River are primarily low-elevation dikes and minimum maintenance practices. Notches and low-elevation modifications have been employed on the middle Mississippi River on approximately 75 dikes (64 notches and 11 low elevations); minimum maintenance practices are also used. On the lower Mississippi River, over 200 dikes have been notched or allowed to remain below design grade for environmental reasons.

An extensive study of aquatic habitat improvement to river training structures was conducted by the Upper Mississippi River Restoration Environmental Management Program in the 1980s and 1990s. This study evaluated the following:

- Closure structures,
- Wing dam notching,
- W-weirs,
- Notched closure structures,
- L-Head dikes,
- Spur dikes,
- Alternating dikes,
- Stepped up dikes,
- Bendway weirs,
- Blunt nosed chevrons,
- Off-bankline revetment,
- Hard points in side channels,

- Vanes,
- Cross vanes and double cross vanes,
- J-hook dikes,
- Multiple roundpoint structures,
- Environmental dredging,
- Longitudinal peak stone toe protection,
- Wood pile structures,
- Root wad revetment,
- Woody debris,
- Boulder clusters, and
- Fish lunkers.

The results and recommendations of this study are contained in an Environmental Design Handbook. This handbook was produced by the US Army Corps of Engineers, Rock Island District (USACE 2006).

10.5 SUMMARY

The use of notches in spur dikes, longitudinal dikes, L-head dikes, and closure dikes is an excellent method for creating varying habitats for fish, birds, and other wildlife. At the same time, dike notches appear to have little to no impact on the efficiency of these types of river training structures to maintain a safe and efficient navigation channel. Other innovative river training structures, such as chevron dikes and hard points, have also shown positive environmental impacts without significantly affecting the navigation channel.

As shown in the UMR Restoration Environmental Management Program, the variety of training structures available to the river engineer is vast and offers great flexibility for environmental sustainment. The emphasis should be to try some type or types of structures that have been successful in similar prototype conditions and monitor the project to help evaluate the overall impact, benefits, and areas of needed refinement.

References

Burch, C. W., P. R. Abell, M. A. Stevens, R. Dolan, B. Dawson, and F. D. Shields Jr. 1984. *Environmental guidelines for dike fields.* Technical Rep. No. E-84-4. Vicksburg, MS: US Army Corps of Engineers, Waterways Experiment Station.

Derrick, D. L., T. J. Pokrefke, M. B. Boyd, J. P. Crutchfield, and R. R. Henderson. 1994. *Design and development of bendway weirs for the Dogtooth Bend Reach, Mississippi River.* Technical Rep. No. HL-94-10. Vicksburg, MS: US Army Corps of Engineers, Waterways Experiment Station.

DuBowy, P. J. 2012. "Environmental benefits of dike notching in the Mississippi River ecosystem." *Inland Port* 2012 (4): 7–11. Accessed November 28, 2018. https://issuu.com/inlandport/docs/inland_port_2012_no.4

Pennington, C. H., F. D. Shields Jr., and J. W. Sjostrom. 1988. "Biological and physical effects of Missouri River spur dike notching." In *Micellaneous Paper EL-88-11*. Vicksburg, MS: US Army Corps of Engineers, Waterways Experiment Station.

Pokrefke, T. J., B. L. McCartney, M. D. Cox, S. W. Ellis, D. C. Gordon, W. H. McAnally, et al. 2014. *Inland navigation: Channel training works*. Reston, VA: ASCE.

Shield, F. D., and M. R. Palermo. 1982. "Assessment of environmental considerations in the design and construction of waterway projects." *Tech. Rep. No. E-82-8*. Vicksburg, MS: US Army Corps of Engineers, Waterways Experiment Station.

USACE. 2006. *Upper Mississippi River System environmental design handbook*. Rock Island District, IL: US Army Engineer District. Accessed February 2, 2016. https://www.mvr.usace.army.mil/Portals/48/docs/Environmental/EMP/EMP_Design_Handbook_August_2006.pdf.

USACE. 2008. *Environmental river engineering project on the Mississippi*. St. Louis, MO: USACE.

USACE. 2012. *Upper Mississippi River restoration environmental management program environmental design handbook*. Rock Island District, IL: USACE.

CHAPTER 11
Fisheries Sustainability

11.1 GENERAL

Maintaining environmental equilibrium on waterways with locks and dams requires provisions for allowing fish migration both upstream and downstream.

Low head locks and dams generally have nonmigrating fish species, which are considered resident fisheries. The life cycle of these fish is confined to freshwater. Normally no fish passage structures are provided at their locks and dams. Examples are the Red River and Tenn–Tom waterways.

On rivers with fish species that migrate between freshwater and saltwater and back to freshwater, special features are required at locks and dams to maintain their survival. The Pacific Coast salmon run has received considerable attention since the building of high head locks and dams on the Columbia and Snake rivers in Oregon and Idaho.

Salmon migration on the Columbia–Snake River system is broken up in the following segments:

- Pacific Ocean to Bonneville Lock and Dam: open river for 145 mi, and
- Bonneville Dam Oregon to Lewiston, Idaho: eight high-lift locks on the Columbia and Snake rivers were built between 1938 (Bonneville Lock and Dam) and 1975 (Lower Granite Lock and Dam)

Initially fish ladders were provided for upstream migration of adult salmon, and downstream juvenile salmon migration was over spillways or through powerhouses. All eight dams have powerhouses. Migratory fish species on the Columbia–Snake Waterway include

- Chinook salmon (adult and jack; jacks are 2-year-old salmon),
- Coho salmon (adult and jack),
- Steelhead,
- Wild steelhead,
- Sockeye salmon,
- Pink salmon,
- Chum salmon,

- Lamprey eels, and
- Shad.

Adult salmon upstream migration is mostly in the spring and fall. Juvenile downstream migration is from late March to fall. Salmon runs in the Columbia were reported to be in the millions in the late 1800s and early 1900s. However, overfishing had depleted the adult chinook runs in the late 1930s when Bonneville Dam was built.

The adult Chinook salmon fish count at Bonneville since construction (1938) follows:

Year	*Average count*
1938–1939	280,000
1940–1949	375,000
1950–1959	360,000
1960–1969	330,000
1970–1979	390,000
1980–1989	300,000
1990–1999	260,000
2000–2009	630,000
2010–2013	820,000

Source: USACE (2016c), Portland District, Fish Counts and Reports.

There is no one factor that can account for these return rates. Factors that affect return rates include floods, habitat, hatcheries, predation, and ocean conditions. Some salmon spend the majority of their life at sea; hence, ocean conditions may be the largest factor in survival rates.

Artificial fish survival methods fall into two categories: upstream adult and downstream juvenile. Upstream methods are

- Fish ladders,
- Trap and transport (trucks),
- Sport and commercial fishing restrictions, and
- Predation control (sea lions).

Downstream methods are

- Transit through powerhouse turbines,
- Transit over spillways (spillway “flip lip” to minimize nitrogen supersaturation),
- Powerhouse bypass,
- Collect and transport (barges or trucks),
- Hatcheries,

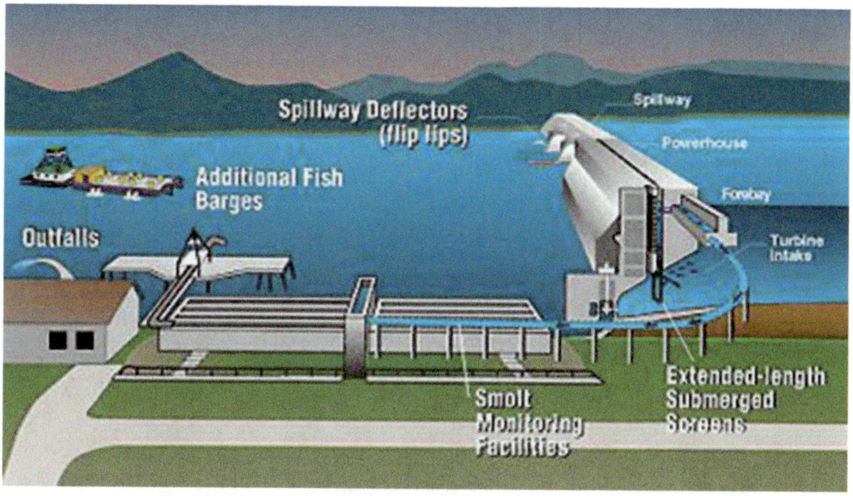

Figure 11-1. Bypass, fish barges, ladders, and other components.
Source: USACE, Portland District.

- Predation control (squawfish, birds), and
- Flow augmentation.

Figure 11-1 shows a generalized view of some passageways at dams.

11.2 FISH LADDERS

11.2.1 General

Fish ladders are structures that enable fish to move on or around barriers to rivers (such as dams, locks, and waterfalls). The ladder is a series of low steps that allow migrating fish to swim or jump to each higher step. The first recorded use of fish ladders was in seventeenth-century France, where bundles of branches were used to create a series of steps in a channel that bypassed a river obstruction. Figure 11-2 shows the main components of a fish ladder.

11.2.2 Chittenden Lock

An early example of construction of a concrete fish ladder in the United States is the Chittenden Locks and Dam in Seattle, Washington. It was constructed in 1917 and included a fish ladder. The locks and dam are part of the Lake Washington Ship Canal Project, which provides a 30 ft deep, 100 ft wide navigation channel into Lake Union and Lake Washington. The ship canal water level is maintained at an elevation 20 to 22 ft above MLLW. This means the

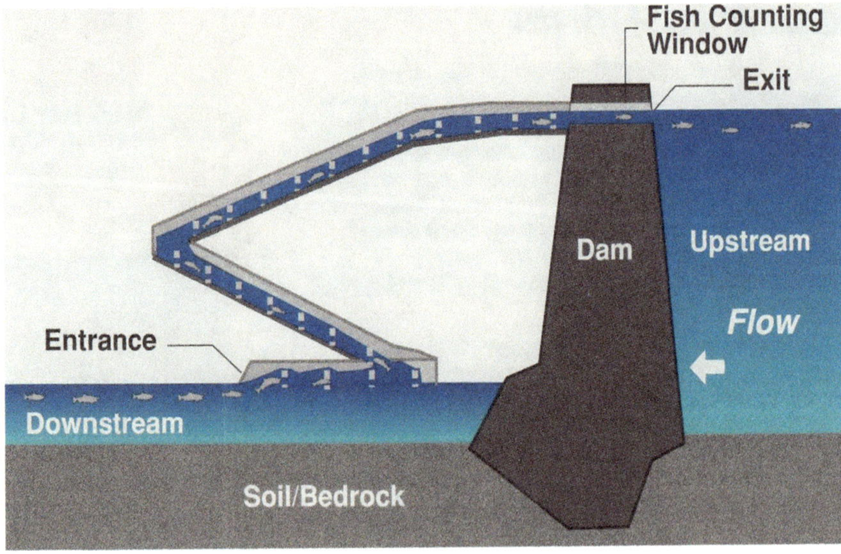

Figure 11-2. Fish ladder.
Source: USACE, Portland District.

difference in water level from the canal to Puget Sound is 6 to 26 ft, depending on the tide level.

The original fish ladder had 10 steps and poor attraction water. The majority of the migrating salmon used the locks to move upstream. Construction of a new ladder with 21 steps provides the fish with a lesser slope than the old 10-step ladder (USACE 1978). The new ladder also gives fish the choice of jumping over weirs or using underwater openings in each weir. The improvements also substantially increased the quantity of attraction water.

The ladder is used by Chinook Salmon, cono, sockeye, steelhead trout and sea-run cutthroat trout. A photo of the fish ladder is shown in Figure 11-3.

11.2.3 Bonneville Dam

Bonneville Dam on the Columbia River is an example of a high-head fish ladder. Migrating salmon must ascend about 70 ft to reach the upper pool. The lock, dam, and powerhouse complex, completed in 1938, had a fish ladder included as a project feature. There are a total of eight dams and fish ladders that allow fish migration from the Pacific Ocean to Lewiston, Idaho. Figure 11-4 shows the Bonneville Dam fish ladder.

11.2.4 John Day Dam

The John Day Dam on the Columbia River has a spillway, powerhouse, navigation lock, fish ladder for adults moving upstream, and a fish bypass system for juveniles moving downstream.

Figure 11-3. Chittenden Lock fish ladder.
Source: USACE, Seattle District.

The navigation lock has a lift of 113 ft. This means that the ladder must accommodate fish vertical transit of 113 ft or more, depending on pool levels. John Day lock is the highest single lift lock in the United States, which makes the ladder the highest in the US inland navigation system. Figure 11-5 shows the John Day fish ladder.

11.3 BYPASS CHANNEL

A bypass channel is another option for fish passage around a river obstruction that would prevent migration. These channels are suitable for low-head barriers. The channel must have water velocities comparable to the natural river flow, so

Figure 11-4. Bonneville Dam fish ladder.
Source: USACE, Portland District.

Figure 11-5. John Day fish ladder.
Source: USACE, Portland District.

the channel slope is usually very flat. A mild channel slope would dictate a very long channel, perhaps kilometers long, for obstruction heads higher than 15 to 20 ft.

An example of the bypass channel concept is at the Lower Yellowstone Dam. The 12 ft-high dam was built in 1905 to store water for irrigation of 54,000 ac of land in Montana and North Dakota. The dam prevented the migration of pallid sturgeon to 165 mi of upstream habitat previously used for spawning and rearing.

Section 7(a)(2) of the ESA requires all federal agencies to use their resources for the conservation and recovery of endangered species and the ecosystem upon which they depend. The pallid sturgeon was listed as endangered in 1990. This prompted the Department of Interior and USACE to develop a recovery plan. The resulting plan consisted of a 11,150 ft (about 2 mi) bypass channel to allow sturgeon access to the upstream habitat. The channel would be 40 ft wide at the bottom and have a top width of 150 to 250 ft. The channel would direct 13% to 15% of the Yellowstone River flow. The channel would have a slope of 0.07% to match the natural river slope of 0.04% to 0.07%. The channel would have an armor layer of large gravel to cobbles, similar in size to the natural river. Estimated cost of this project was $59 million. The bypass channel is recommended in an April 2015 final supplement to the Final Environment Assessment.

11.4 TRAP AND TRANSPORT

The trap and transport of adult salmon migration upstream is an option for dams that have high heads. Migration fish are attracted to a collection facility below the dam. The fish are trapped and transported by trucks for release above the dam, so they can proceed to their spawning grounds, or the fish are taken to a hatchery for production of juvenile salmon.

An example of this trap and transport operation is the Cougar Dam on the McKenzie River (tributary of the Willamette River, Oregon), a 452 ft high rockfill dam. The collection facility includes a fish ladder leading from the base of the dam to a fish collection and sorting area. From there, adult salmon, bull trout, and other resident fish species are loaded into trucks and transported to release locations above Cougar Reservoir. Biologists estimate that the habitat above the dam once supported more than 4,000 returning adult spring Chinook. The trap and transport operation initiated in 2010 may reach several hundreds of fish (USACE 2016c). Figure 11-6 shows the Cougar Dam.

11.5 FISHING RESTRICTIONS

Fishing regulations and restrictions are another method to manage the fisheries in inland waterways. These regulations are developed and enforced by state fish and wildlife agencies. In some cases, regulations are a cooperative effort by state agencies

Figure 11-6. Cougar Dam.
Source: USACE, Portland District.

and Native American tribes. Such is the case of regulations on the Columbia where the fishing is comanaged by the Oregon and Washington Departments of Fish and Wildlife and four treaty tribes, represented by the Columbia River Inter-Tribal Fish Commission. An example of these regulations is the Oregon state 2015 fishing regulations. Some of these restrictions follow:

- Minimum and maximum fish size that can be kept,
- Number of fish allowed per day,
- Barbless hooks only,
- Keep only hatchery-raised fish. Hatchery fish are identified by removal of adipose fin,
- Dates (seasons) when fishing is allowed,
- Anglers are required to have a valid license,
- Restricted fishing areas (usually a specified distance below dams), and
- Commercial fishing is not allowed in designated portions of a river.

Often Native American tribes are exempt from some of these regulations.

11.6 PREDATION CONTROL FOR ADULTS

There are many predators of adult salmon, such as killer whales, bears, sea mammals, and people. Human predation is controlled by fishing regulations, and most of the other predators are allowed to pursue natural patterns. One exception is the West Coast sea lions. The California and Steller sea lions have roamed the Pacific coast for centuries. They range from Baja, California, to southeast Alaska. California sea lions are large marine mammals, with males measuring up to 8 ft long and weighing 1,000 lb or more. Females can reach 6 ft long and weight 250 lb. Steller sea lions are larger.

Since the early 1980s, California sea lions have been moving in increasing numbers up the Columbia River, first to the Astoria area (near the mouth), then to Bonneville Dam, 145 mi from the mouth. At Bonneville Dam, the estimated salmon predation by sea lions is 4,000 to 6,000 per year for 2008 to 2010. This constitutes about 2% to 3% of the salmon runs passing the Bonneville Dam project.

In addition to salmon predation, white sturgeon are also on the sea lion diet. The estimated sturgeon loss has increased from 312 in 2006 to 1,879 in 2010. This translates to 25% of the observed catch before 2008 to 16% between 2008 and 2010 (Stansell et al. 2011).

Like all marine mammals, California sea lions are managed under the federal MMPA of 1972. They are designated as "depleted" under MMPA and not listed for protection under the ESA. The MMPA law was amended in 1994 to provide a process for states to lethally remove individual California sea lions that threaten recovery of salmon and steelhead stocks listed under ESA. This provision does not apply to Steller sea lions, which are listed as "threatened" under ESA.

Methods to control California sea lion predation have been hazing, relocation, and euthanasia. Hazing by noise had no effect on sea lions' behavior. Relocation by trap and transport to a remote location did not work because they often returned to the site where they were captured. However, some were placed in zoos and aquariums. Some were euthanized by lethal injection. From 2008 to 2012, 54 California sea lions were removed from the Columbia River. Thirty-eight were euthanized, 11 were placed in zoos and acquariums, and five died during capture activities.

11.7 TURBINE BYPASS MEASURES

The juvenile fish bypass systems in place at the lower Columbia–Snake River dams guide fish away from turbines by means of submerged screens positioned in front of the turbines. The juvenile fish are directed up into a gate well, where they pass through orifices into channels that run the length of the dam. The fish are then routed back out to the river below the dam or, at the four dams with fish transport facilities, fish can be routed to a holding area for loading on specially equipped barges or trucks for transport downriver.

The juvenile bypass systems guide 80% to 90% of steelhead salmon and 60% to 70% of spring/summer Chinook salmon away from the turbines and upward through the bypass channel. This percentage measure is called *fish guidance efficiency*, and the rates vary from dam to dam. This bypass channel is shown in Figure 11-7, at Little Goose Dam on the Snake River. The bypass channel is above the fish ladder (USACE 2016a).

Figure 11-7. Little Goose Dam fish bypass.
Source: McCartney et al. (1998).

The discharge for the bypass can be directly below the spillway or at a location farther downstream. A near dam discharge is usually a very turbulent area where predators (squawfish and birds) are a concern. However, some bypass discharges are a considerable distance downstream to a section having fast-moving currents. An example is the Bonneville second powerhouse, where the discharge is about 1.5 mi downstream of the dam.

11.8 JUVENILE PASSAGE THROUGH AND AROUND POWERHOUSES

11.8.1 Passage Through Turbines

Turbine blades move at the same speed as water flowing through the turbines, just as a person walking through a revolving door moving at the same speed as the door rotation. Damage to fish occurs not by cutting but when they experience negative pressures near the turbine blade. Newer turbines reduce or eliminate this negative pressure, reduce fish mortality, and are considered fish friendly.

Survival rates of fish passing through turbines have been estimated at the eight Columbia/Snake Rivers for several years. These estimates are made using field test measurements and computer analysis. A sample of these estimates is shown here. This sample is for yearling Chinook salmon for 2010 and 2011.

Project	*Turbine Survival for the Average 2010–2011*
Bonneville, first powerhouse	97%
Bonneville, second powerhouse	95%
The Dalles	90%
John Day	85%

Source: Ploskey et al. (2012).

11.8.2 Passage Over Spillways

Problems associated with discharging water over spillways occur when air is entrained in the water as it plunges into the spillway basins. This raised level of gas supersaturation can be harmful to fish. Spillway deflectors have been installed at five of the eight lower Columbia and Snake dams to produce a more horizontal spill flow and limit the plunge depth of water over the dam spillway, reducing the amount of entrained nitrogen. Installation of spillway deflectors has begun at two more dams. These deflectors are called a "flip lip." Figure 11-8 shows this flip lip.

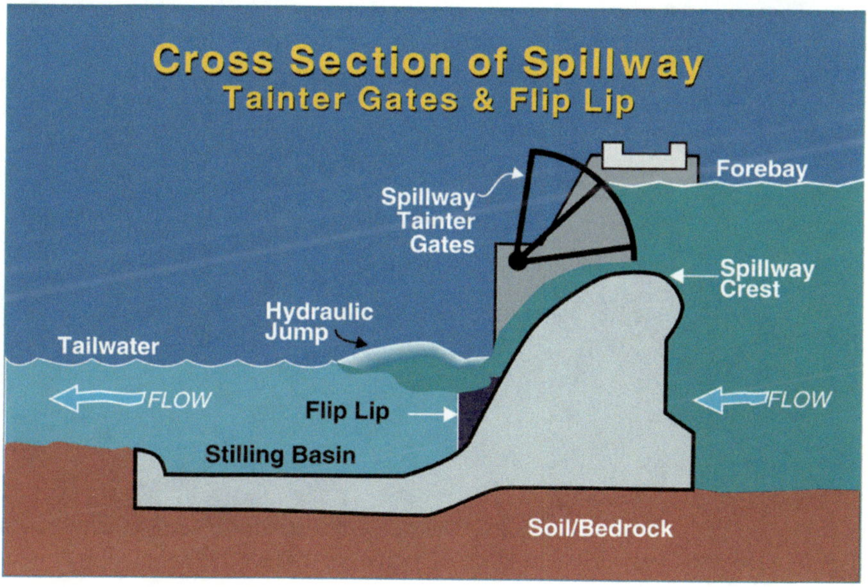

Figure 11-8. Spillway flip lip.
Source: USACE, Portland District.

11.9 JUVENILE FISH TRANSPORT

Three of the four Snake River dams and McNary Dam on the Columbia River have fish transport facilities. At these four dams, juvenile fish that go through the bypass systems can be routed either directly back into the river below the dam or to holding and loading facilities for loading into barges or trucks for transport. The transport barges and trucks carry the fish past the remaining projects for release below Bonneville Dam. River water circulates through the barges, allowing the fish to imprint the chemicals and smells of the water during the trip downriver. The barges have a closed-circuit recirculation system, which can shut off water intake in case of contamination in the river. They also have pumping systems, which can help de-gas the water in areas where gas supersaturation is a problem.

The COE runs the Juvenile Fish Transportation Program in cooperation with National Marine Fisheries Service, and in accordance with the National Marine Fisheries Service Hydropower Biological Opinion for salmon. Fifteen to 20 million salmon and steelhead have typically been transported each year over the past several years (USACE–Northwestern Division, n.d.). A fish transport barge is shown in Figure 11-9.

Figure 11-9. Fish transport barge on the Columbia River.
Source: USACE, Portland District.

11.10 HATCHERIES

Fish hatcheries have been used for many years as a means of supplementing the natural fish stock. There are 70 national hatchers administered by the USFWS. These hatcheries are located in most states in all regions of the United States.

In addition to the NFHs, there are numerous hatcheries operated by the state fish and wildlife agencies. Also, some hatcheries are operated by Native American tribes.

The intent of hatcheries is to compensate for declining fish populations caused by

- Overfishing,
- Climate change (water temperature),
- Barriers to migrating fish species,
- Pollution, and
- Lost habitat.

Hatcheries can be viewed as producing a commodity for sport and commercial fishing and not a true environmental sustainable activity. However, they can also be considered an environmentally sustainable practice.

11.11 PREDATION ON JUVENILE FISH

11.11.1 General

Predation of juveniles can be from both large fish and birds. Two examples presented here are the double-crested cormorant, a migratory seabird, and the northern pike minnow. Other predators include but are not limited to Caspian terns, gulls, small-mouth bass, and walleye. All feed on juvenile salmon on their journey to the ocean.

11.11.2 Double-Crested Cormorant

The double-crested cormorant is a long-necked, black seabird that established a breeding colony near the mouth of the Columbia River. This island (East Sand Island) was created from dredge material taken from the adjacent Columbia River ship channel.

The low sand island proved to be an excellent breeding habitat for the cormorant, and 100 nesting pairs in 1989 grew to about 15,000 pairs in 2013. The double-crested cormorant is a federally protected migratory bird. However, their main food source is juvenile salmon headed to the ocean. It is estimated that 11 million of these salmon are consumed by the cormorant each year. Figure 11-10 shows cormorants on East Sand Island.

Figure 11-10. Nesting cormorants.
Source: USACE, Portland District.

A plan to control the cormorant colony was developed in 2014. The plan is to reduce the land area suitable for nesting sites. This will reduce the bird population but not eliminate it all together (USACE – Portland District, n.d.).

11.11.3 Northern Pike Minnow

The northern pike minnow fish is a large member of the minnow family. They can live longer than 15 years, reaching 24 in. and 8 lb. The pike minnow are voracious predators, and in the Columbia and Snake rivers, salmon smolts (juveniles) compose a large part of their diet. The hydropower projects on the Columbia and Snake rivers have provided an excellent habitat, and pike minnows have thrived.

The negative impact of pike minnows on the juvenile salmon migration prompted a control program by paying fishermen for pike minnows longer than 9 in. Rewards range from $4 to $8 per fish. The goal of the program is not to eliminate the pike minnow population but to reduce the average size and curtail the number of older fish. Since 1990, this reduction program has removed more than 3.5 million pike minnows from the Columbia and Snake rivers. This reduces pike minnow predation on young salmon by about 4 to 6 million per year (BPA 2010).

11.12 WATER MANAGEMENT

11.12.1 General

There are two categories of inland waterways projects that control water flow. These are run-of-the-river and storage dams. The run-of-the river dams have very little storage capacity and therefore very little discharge control. The dam's only purpose is to maintain low velocity impoundments with adequate depth for movement of barges or ships. These run-of-the-river projects are usually low head,

30 ft or less, and have no flood control function; however, there are some high-head locks and dams (like the Columbia–Snake system) with heads of up to 110 ft. These projects often include powerhouses to harness the flow that does not go through the locks or over a spillway. The Bonneville project on the Columbia River is this type of project. Bonneville has two powerhouses, a spillway, and navigation lock. The normal head on this project is 69 ft.

The second category is storage dams. These dams capture seasonal rainfall and snowmelt for release at a later time. The projects are used for flood control and irrigation, and they may or may not have a hydropower component. Storage dams can be very large, such as Grand Coulee and Chief Joseph dams on the Columbia River, or they can be rather small like many dams of the Willamette River tributaries. Control of river flow for fish survival is of much greater concern for storage dams.

Upstream storage dams hold back water for flood damage reduction and other uses, interrupting the seasonal river flow patterns. Seasonal releases of water from the dams, called *flow augmentation*, can aid salmon migration. The COE operates two upstream storage dams—Dworshak Dam in Idaho and Libby Dam in Montana—which are used in flow augmentation for migrating juvenile salmon in accordance with the National Marine Fisheries Service Hydropower Biological Opinion. Water is released in spring and summer months to improve flows in the Snake and Columbia rivers. The COE coordinates flow augmentation with the region through a Technical Management Team, which meets weekly during the juvenile fish migration seasons to discuss flows and spills and to plan operations for fish.

11.12.2 Run-of-the-River Dams

Run-of-the river dams are typically operated via spillway gates to maintain a rather narrow range of water levels in the pool above the dam to facilitate navigation. Flood damage reduction is not a primary consideration, although such dams are typically operated so as to avoid making flooding worse. A recent trend for the run-of-the river spillway gate operation is to abolish the practice of a "hinge pool" scenario. The hinge pool allows a lowering of the water levels of dams by using lower spillway gate openings to cause a slope on the water surface near the dam. This lower water surface will reduce side channel flooding, which in turn decreases aquatic habitat. The hinge pool concept is discussed further in the Upper Mississippi case history.

11.12.3 Large Storage Dams

Columbia River

The Columbia River is longer than 1,200 mi. It has its headwaters in Canada, flows through Washington State, and empties into the Pacific Ocean at the Oregon/Washington Coast. The Columbia system and tributaries have more than

60 major dams, with some having considerable storage capacity. The Grand Coulee Dam at 550 ft tall and Chief Joseph Dam at 236 ft tall are two of these large storage dams that capture high flood flows for release during low flow periods. This operation reduces downstream flooding and stores water for summer irrigation use. These storage dam releases have also been used to speed flow velocities to reduce the transit time for juvenile fish migrating to the ocean.

Missouri River

The Missouri River mainstream system is 2,321 mi, starting in Montana and terminating at the Mississippi River, near St. Louis, Missouri. The upstream portion of the river has six high-head storage dams: Fort Peck, Garrison, Oahe, Big Bend, Fort Randall, and Gavins Point (Gavins Point is the farthest downstream dam). The lower portion of the river includes miles of open river (no dams) navigation channel. Figure 11-11 shows the Gavins Point Dam.

Prior to the construction of the dams on the Missouri River, two major increases in flow occurred each spring, one due to the melting of plains snowpack and the second due to the melting of mountain snowpack and spring rains throughout the Missouri River Basin. Construction and operation of the six large dams on the mainstem of the Missouri River have altered the historical spring and early summer high flows. Flood peaks have been greatly reduced, and the stored water is released at appropriate levels to serve the downstream uses of flood control, navigation, water supply, hydropower generation, and recreation. These releases have been altered to also benefit water quality and fish and wildlife

Figure 11-11. Gavins Point Dam.

resources, including the endangered interior least tern and threatened piping plover. In the meantime, the benefits of the higher spring and early summer flows to the native river fish were lost.

In response to the needs of the native river fish on the Lower Mississippi River (LMR), especially the endangered pallid sturgeon, March and May spring pulses from Gavins Point Dam were included in the Missouri River Mainstem Reservoir System's Master Water Control Manual in March 2006. Spring pulses created spring rises on the downstream Missouri River in May 2006 and March 2008. These pulses were released to mimic the much larger, historic spring rises on the Missouri River, which still occur naturally as one proceeds farther down the LMR. The benefits of the created spring rises are, therefore, to be more significant in the upper 200 mi of the Lower Missouri River from Gavins Point Dam to about Omaha, Nebraska. Spring 2016 was the scheduled completion date for the evaluation of the pulse strategy (USACE 2016d).

11.12.4 Small Storage Dams

Small storage dams are those located on tributaries of mainstream rivers. These dams are primarily used for flood control by release during the dry season. These dams can be very high, and many have small hydropower plants. A problem with deep reservoirs is the cold-water discharge from the bottom of the reservoirs through outlet works. A remedy for this problem is to have an intake tower with a capability of drawing water from different levels of the reservoir. This type of structure is called a *multilevel intake.* Figure 11-12 shows three types of multilevel intakes. An example of a free-standing intake tower is Cougar Dam on the south fork of McKenzie River near Eugene, Oregon. This flood control dam reduces flood peaks on the Willamette River. The following are the dam characteristics:

- Constructed in 1963,
- Length of 1,600 ft,
- Height of 452 ft,
- Type: rock fill, gated concrete spillway,
- Reservoir length of 6 mi, and
- Powerhouse: two 25 MW generators.

To correct the water temperature control problem, the intake tower was modified to add multilevel ports in 2005. This new multilevel intake tower draws water from different depths within the reservoir, mixing it to a temperature that more closely replicates prereservoir downstream temperatures. Figure 11-13 shows the intake.

11.12.5 Reregulating Dams

A dam that generates electric power is often operated during peak power demand periods. This means that discharges from this type of dam can have daily and weekly fluctuations that can adversely affect downstream fisheries. A method to

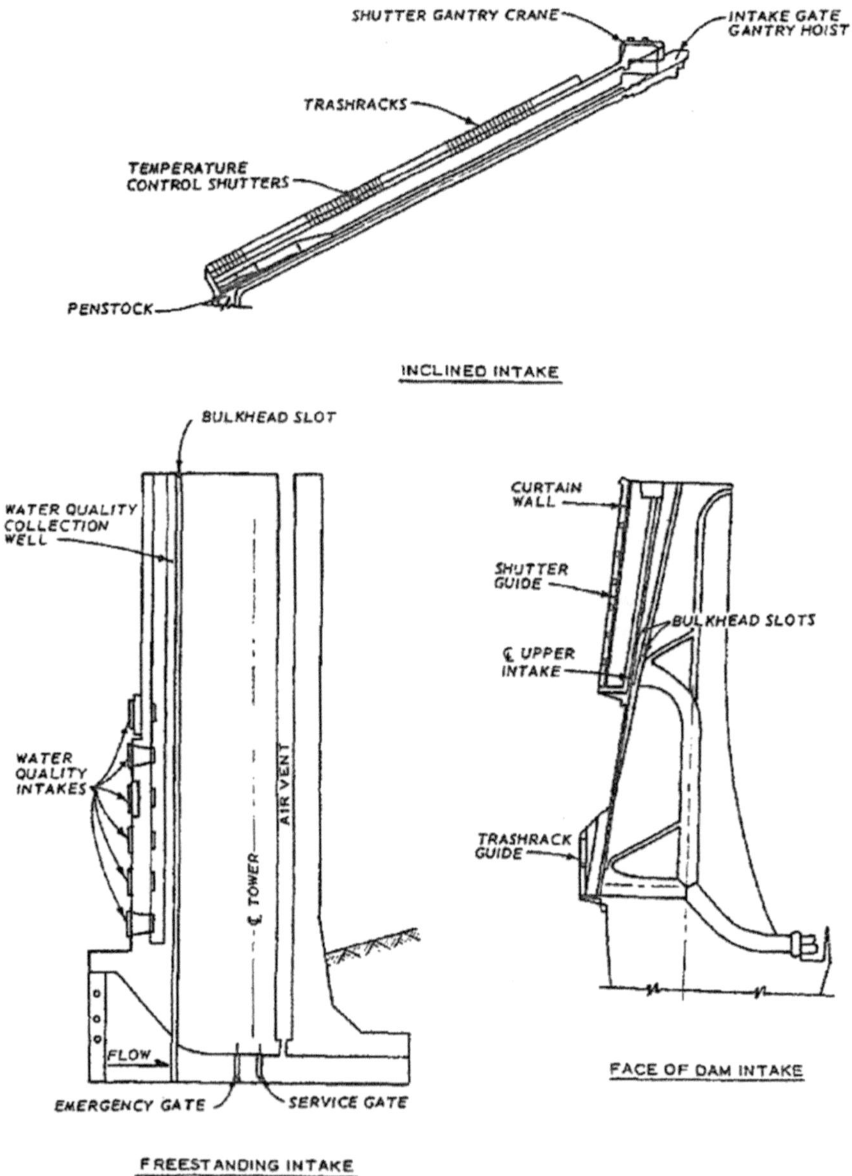

Figure 11-12. Water quality intake types.
Source: USACE (1980).

reduce these large fluctuations is to build a reregulating dam downstream from the primary dam. These reregulating dams will store the peak inflow and release a more uniform flow downstream.

Figure 11-13. Cougar intake tower.
Source: Courtesy Scott Annus, USACE, Portland District.

An example of a reregulating dam is the Foster Dam located on the Santiam River, Oregon. The Santiam River is a tributary of the Willamette River. Foster Dam is about 5 mi downstream from Green Peter Dam, and the two dams were completed in 1968. Green Peter Dam (upstream dam) is 327 ft high, has two generators at 80 MW, and has a 10 mi long reservoir. Foster Reregulation Dam is 126 ft high, has two generators 20 MW, and has a 3.5 mi long reservoir. In addition

Figure 11-14. Foster Dam.
Source: USACE, Portland District.

to the dam's reregulation function, the project has a fish hatchery and fish collection facility. Figure 11-14 shows a photo of Foster Dam.

References

BPA (Bonneville Power Administration). 2010. *Predation control helps salmon.* Portland, OR: Bonneville Power Administration. Accessed January 26, 2016. http://www.salmonrecovery.gov/files/factsheets/predatorcontrol-sept2010.pdf.

McCartney, B. L., J. George, B. K. Lee, M. Lindgren, and F. Neilson, eds. 1998. *Inland navigation: Locks, dams, and channels.* Reston, VA: ASCE.

Ploskey, G. R., M. A. Weiland, and T. J. Carlson. 2012. *Route-specific passage proportions and survival rates for fish passage through the John Day Dam, The Dalles Dam, and Bonneville Dam in 2010 and 2011.* Richland, WA: Pacific Northwest National Laboratory.

Stansell, R. J., K. M. Gibbons, W. T. Nagy, and B. K. van der Leeuw. 2011. *Evaluation of pinniped predation on adult salmonids and other fish in the Bonneville Dam Tailrace: 2011 field report.* Portland, OR: Fisheries Field Unit, Bonneville Lock and Dam.

USACE. 1978. *Lake Washington ship canal fish ladder.* Seattle WA: US Army Engineer District.

USACE. 2016a. "Columbia River fish mitigation." *Fact sheet.* Northwestern: US Army Engineer Division. Accessed February 2, 2016. https://www.nwd.usace.army.mil/media/fish-migration.aspx.

USACE. 2016b. "Cougar Dam adult fish collection facility." *Fact sheet*. Portland, OR: US Army Engineer District. Accessed March 26, 2015. https://www.nwp.usace.army.mil/media/factsheetarticleview/tabid/2043/article/492592/cougar-dam-adult-fish-collection-facility.

USACE. 2016c. "Fish counts and reports." *Fact sheet*. Portland, OR: US Army Engineer District. Accessed February 2, 2016. https://www.nwp.usace.army.mil/missions/environmentalfish/count.aspx.

USACE. 2016d. "Flow modification, Missouri River recovery program." *Fact sheet*: US Army Corps of Engineers. Accessed February 8, 2016. http://moriverrecovery.usace.army.mil/mrrp.

USACE. n.d. "Columbia River estuary cormorants environmental impact statement." *Fact sheet*. Portland, OR: US Army Engineer District. Accessed February 17, 2015. http://www.nwpusace.army.mil/missions/current/cormorantEIS.aspx.

USACE. n.d. *Environmental engineering for shallow draft waterways*. Washington, DC: USACE.

USACE and US Department of Interior. 2015. *Intake diversion dam modification, Lower Yellowstone Project, Montana*. Washington, DC: USACE and US Dept. of Interior.

WDFW (Washington Department of Fish and Wildlife). n.d. *Columbia River sea lion management*. Accessed July 3, 2016. http://wcpw.wa.gov/conservation/sealions/questions.html.

CHAPTER 12
Other Sustainability Considerations

12.1 MIGRATORY BIRDS

There are three main migratory flyways in the continental United States—the Atlantic coast, Mississippi River, and Pacific coast. The Atlanta and Pacific routes are primarily along coastal estuaries and freshwater tributaries that feed these estuaries. Inland waterway development has had little impact on these migration routes.

The development of the Mississippi River and tributaries for navigation and flood control has entailed converting from a shallow, braided, meandering system to a confined system with deeper depth, reduced meanders, (bendway, oxbows, etc.), and level channels. What had been large seasonal floodplains has been converted to farmland.

The loss of floodplains impacted migratory birds by reducing their resting and feeding areas needed on their migration route. Any environmental mitigation or restoration will need to address this loss of habitat for migrating birds and provide suitable replacement land to support the sustainable waterfowl population.

12.2 LAND MANAGEMENT

Most new or restoration projects include land acquisition to mitigate for land lost to a waterway-created reservoir. This "dry" land purpose is to provide habitat for land dwellers and birds. These animals can be both residents and migratory.

Wildlife management plans and enforcement are needed to ensure that the reproduction habitat requirements are met for the target species. Explanation of management methods and examples are beyond the scope of this technical report; however, information on this subject is available through state and federal fish and wildlife agencies.

12.3 SALINITY CONTROL

The Lake Washington ship channel was constructed in 1917 to provide a freshwater ship channel from Puget Sound (saltwater) to Lake Union and Lake Washington (freshwater). Figure 12-1 shows the Chittenden Lock, which is at the downstream of the project.

The canal that connects the navigation locks to Lake Union then on to Lake Washington is 30 ft deep and 100 to 300 ft wide. The Chittenden Locks consist of a large ship lock chamber 80 × 825 ft and a 36 ft depth. The small lock is 30 × 150 with a depth of 16 ft. Depth refers to the mean lower low water levels in Puget Sound, so saltwater enters the locks when the lower gates are opened.

To retard the advance of saltwater into Lake Washington, a sump or saltwater basin 2,000 ft long × 200 ft wide and 20 ft deeper than the navigation channel was excavated immediately upstream from the main lock. The saltwater basin is drained with a conduit to the downstream face of the dam spillway. This drain removes the denser saltwater on the bottom of the saltwater basin back to Puget Sound. This drain project was completed in 1934 (USACE 1971).

Other methods to reduce saltwater intrusion are to use a small lock whenever possible. Currently, the majority of vessels using the locks are pleasure crafts or fishing boats. Also, a movable barrier wall was installed on the large lock. The wall (steel gate) lies flat on the lock floor and is raised when small boats are being

Figure 12-1. Chittenden Lock looking toward Puget Sound.
Source: USACE, Seattle District.

locked. This acts as a barrier to saltwater entering the chamber upstream from the wall. The wall is lowered only for the occasional deep-depth vessel transit.

12.4 INVASIVE SPECIES

The Nonindigenous Aquatic Nuisance Prevention and Control Act of 1990 defines an aquatic nuisance species as a nonindigenous species that threatens

- Diversity or abundance of native species,
- Ecological stability of infested water, or
- Commercial, agricultural, aquacultural or recreational activities dependent on such waters.

Asian carp are one of these species that has invaded the Mississippi River system. These carp are voracious eaters; an adult carp can eat half its body weight each day. The Asian bighead carp can grow to more than 100 lb, with an average size of 40 lb. Silver carp, called "jumping carp," may land in boats, damage property, and injure people (USACE 2012).

A concern over the invasion of carp in the UMR has caused the 2015 closure of the upper St. Anthony Falls lock near Minneapolis, Minnesota. The lock was opened in 1963 with a normal lift of 49 ft and provided access to the Mississippi River and St. Croix River above Minneapolis. The closure was mandated by the Water Resources Reform and Development Act of 2014 to prevent the invasion of Asian carp above St. Anthony Falls Lock and Dam. This lock was closed in June 2015, and this was the first time a navigable waterway has been closed to prevent migration of an invasive carp. The lock will be maintained so that it can be used for flood control (USACE 2015). Figure 12-2 shows the upper St. Anthony Falls Lock.

Another non-native fish posing an ecological problem is the northern snakehead fish. This freshwater fish can thrive in cold water and can grow up to 4 ft long. It is considered established in several East Coast states (e.g., Virginia, Maryland, New York) and collected in Florida, Illinois, and North Carolina.

12.5 AQUATIC PLANT CONTROL

Should shallow draft channels become infested with aquatic plants to the point that navigation or any other desired use of the waterway is affected, several aquatic plant management measures are available for bringing those infestations under some degree of control. The control technique to be used is dependent on the species of aquatic plant causing the problem, its magnitude, its location, and the site characteristics of the channel itself. The degree of control that will bring the problem to an acceptable level must also be a consideration.

Aquatic plants that can cause problems in shallow draft channels are of two basic types—floating plants, such as water hyacinth, and submersed plants, such as

Figure 12-2. Upper St. Anthony Falls Lock.
Source: USACE, St. Paul District.

hydrilla or Eurasian water milfoil, which grow up from the channel substrate. The three major technology areas of aquatic plant control that might be used in shallow draft channels are biological, chemical, and mechanical controls, or some combination of two or more of these methods.

12.5.1 Biological Control

Biological control is the use of living organisms for controlling other living organisms. The control organism may feed on the target organism or affect it in some other way, such as to reduce its numbers or growth. Biological control agents potentially available for use in aquatic plant control are insects for control of alligator weed and water hyacinth, plant pathogens for control of water hyacinth, and herbivorous fish for control of submersed species. In general, biological methods are relatively inexpensive but take considerable time to become effective.

12.5.2 Chemical Control

The application of safe and effective chemical agents for aquatic plant control is a proven method available for use in shallow draft channels. Approved chemical agents for aquatic use may be liquids that can be sprayed onto floating plants or inserted under the water for controlling submersed plants, or they may be solids that can be applied by spreaders, for example, over the surface of the channel. Chemical methods are generally readily available and are relatively inexpensive when compared to other methods.

12.5.3 Mechanical Control

Mechanical devices for controlling aquatic plants vary from deflecting booms and screens or clipping bars mounted on boats to more sophisticated systems whereby the plants are cut and removed from the water to disposal areas. Mechanical methods are generally rather costly but sometimes desired over other methods because no outside agents are added to the environment, such as organisms or chemicals. This may be particularly true in politically sensitive areas (USACE n.d.).

12.6 HISTORIC/ARCHAEOLOGY

Historic preservation is the act of identification, evaluation, recordation, documentation, curation, acquisition, protection, management, rehabilitation, restoration, stabilization, maintenance, research, interpretation, conservation, and education and training for cultural, built, and/or engineered environments (USACE 2000).

In 1966, the National Historic Preservation Act sought to protect, restore, and maintain historical and archaeological resources affected by federal projects. This legislation created a federal–state partnership to identify districts, sites, objects, buildings, and structures significant in American history, archaeology, and culture. It also established the Advisory Council on Historic Preservation, requiring federal agencies that had direct and indirect jurisdiction over proposed federal projects to take into account the effect of those projects on cultural resources eligible for listing in the National Register of Historic Places. Congress provided funding for cultural resource management projects through the Archaeological and Historic Preservation Act of 1974, which granted federal agencies the authority to devote up to 1% of a project's total construction cost to archaeology.

Legislation in 1979 further expanded the federal government's role in evaluating and protecting cultural resources. That year, the Archaeological Resources Protection Act established a permit procedure for investigations of archaeological resources on public lands, prohibiting the removal, sale, receipt, and interstate transportation of these resources obtained without a permit from public or Indian lands. This legislation ensured that individuals and organizations wishing to investigate or excavate and remove archaeological resources from federal lands had the necessary professional qualifications and that federal guidelines for research and curation were followed. Congress also passed the Native American Graves Protection and Repatriation Act in 1990, in response to Native Americans' concern about the loss of human remains and cultural items. This legislation directed federal agencies to inventory their collections of human remains and associated funerary objects and to identify the descendants entitled to claim them (USACE 2003).

An example of this historic/preservation effort is the Bonneville Dam second powerhouse, which was completed in 1981. This project is part of the lock, dam, and powerhouse complex located 145 mi from the mouth of the Columbia River and 40 mi east of Portland, Oregon. The project consists of a navigation lock, spillway, and two powerhouses.

During initial studies for a second powerhouse, an abandoned Native American village was identified in the proposed construction area.

This village was first visited by Lewis and Clark on their expedition, 1805–1806. The village was unique because of its size and location. Situated on the Columbia River, it was a major fishing village on the Columbia River trade route.

Prior to powerhouse construction, archaeology teams excavated the site and unearthed more than 600,000 artifacts. These artifacts were preserved and cataloged and then stored at Bonneville. At their request, the artifacts were transferred to the Yakama Nation for preservation. A description of this effort is contained in Chapter 13, Section 13.5 (USACE 2003).

References

USACE. 1971. "Estuarine navigation projects." *Tech. Bull. No. 17*, Committee on Tidal Hydraulics. Vicksburg, MS: USACE.

USACE. 2000. *Planning guidance notebook*. ER 1105-2-100. Washington, DC: USACE.

USACE. 2003. *Currents of change: A history of the Portland District, US Army Corps of Engineers 1980–2000*. Portland, OR: US Army Engineer District.

USACE. 2012. *Asian carp and aquatic nuisance species*. Chicago: US Army Engineer District.

USACE. 2015. *Upper St. Anthony Falls lock closure*. St. Paul, MN: US Army Engineer District.

USACE. n.d. *Environmental engineering for shallow draft waterways*. Washington, DC: USACE.

CHAPTER 13
Case Studies

13.1 GENERAL

Several examples of environmentally sustainable practices are provided to give more detail of discussions given in the text. The following case studies are discussed:

- Red River Waterway: EIS,
- UMR: Restoration,
- Lock 27 Mississippi: Major rehab EA,
- Bonneville Dam Second Powerhouse: archeological investigation example,
- Columbia River Deepening Project: EIS and AM,
- Great Lakes: Rehabilitation,
- LMR: Restoration, and
- Willamette River: Restoration of "Oregon chub."

13.2 RED RIVER

The Red River Waterway (J. Bennett Johnston Waterway) is located in Louisiana and Texas. Until the early 1930s the Red River was a tributary of the Mississippi. During 1931, a cutoff of the Mississippi was made, which resulted in the diversion of the Red River to the Atchafalaya River. The Red/Atchafalaya are now connected to the Mississippi by the old river control complex. The Red River project provides a navigation channel from Shreveport, Louisiana, to Daingerfield, Texas.

The authorized head of navigation is located at preproject River Mile 284.5. After the river was shortened by channel realignments, the postproject river mile is 234.4. This project includes five locks and dams, which were completed by 1994.

13.2.1 First Environmental Statement

The final EIS for the Red River Waterway project was completed on April 2, 1973. This EIS discussed the impact to the following:

- Land: channel realignment, water inundation caused by pools, and dredged material disposal sites,

- Vegetation: channel realignment and 78 bendway cutoffs,
- Wildlife: loss of land to game and furbearers,
- Nongame wildlife: loss of habitat to birds, mammals, amphibians, and reptiles,
- Agriculture: loss of land to channel realignment, dredge material disposal,
- Surface water: impact of converting free-flowing water to a series of pools,
- Groundwater: impact on groundwater of inundation of pools,
- Fisheries: bank stabilization reduce off channel fisheries habitat; impact of maintenance dredging,
- Recreational esthetics: impact on boating picnicking, walking, etc.,
- Archeological and historic: reviewed historic/archeological information and will present any significant finds during design or construction, and
- Existing development: impact of modification to bridges, pipelines, power transmission lines, and communication lines.

The following remedial, protective, and mitigation measures are recommended by this EIS:

- Closure plugs installed in the upper end of 39 bendways. These measures will prevent filling of the oxbows to provide fisheries habitat and recreational opportunities. Project design criteria called for preserving oxbows that were at least 1 mi in length. Construction was for a nonovertopping earth closure dam across the upstream end of the oxbow and dredging the lower end of the oxbow if needed to maintain conductivity to the main channel,
- Placement of dredged material (initial construction and maintenance) in locations that do not adversely affect adjacent waters,
- Project contractors exercise care in handling and storage of hazardous material,
- Esthetical features incorporated in the final plan, and
- Prevent adverse impact on the archeological or historic element located in the project area. (USACE 1973)

13.2.2 Second Environmental Statement

A second EIS for the Red River waters was produced in September 1983. This EIS was needed to address a June 1980 Congress authorized the establishment of the Tensas National Refuge. The EIS focus was to mitigate for environmental losses considering the new status as a national refuge on project lands.

The reevaluation of project impacts concluded that the refuge alone will not mitigate for the loss or reduction in quality of about 28,000 ac of terrestrial wildlife habitat.

The selected mitigation alternative consists of acquisition and management of 14,000 ac of forested wetlands, woodlands, and open land. The mitigation lands purchased would be spread over a broad area of the Red River Valley (USACE 1983).

13.2.3 Additional Environmental Features

Hinged crest gates were included at L&D 4 and L&D 5 to improve dissolved oxygen downstream.

13.3 UPPER MISSISSIPPI AND ILLINOIS RIVER SYSTEM RESTORATION

13.3.1 History

Prior to widespread European settlement of the region, the UMRS was a diverse landscape of tall grass prairie, wetlands, savannas, and forests. Logging, agriculture, and urban development over the past 150 years have resulted in the present floodplain landscape that is more than 80% developed. Millions of acres of wetland drainage, thousands of miles of field tiles, road ditches, channelized streams, and urban storm water sewers accelerated runoff to the main stem rivers. The modern hydrologic regime is highly modified, with increased frequency and amplitude of changes in river discharge. Dams and river regulation throughout the basin also modify river flows. The modern basin landscape delivers large amounts of sediment, nutrients, and contaminants to the river. Since impoundment, sediment accumulation and littoral (i.e., wind and wave) processes in the navigation pools have greatly altered aquatic habitats.

At the historic systemwide scale, there were natural gradients in habitat among river reaches. Northern river reaches were more forested and comprised mixed silver maple forests, river channels, seasonally flooded backwaters, floodplain lakes, marsh, and prairie. Beginning around the northern Iowa border and along the lower Illinois River, grasslands and oak savanna dominated floodplain plant communities. Historic surveys reveal a higher proportion of oaks and other mast trees in the forest community than at present. Below the Kaskaskia River, the floodplain was heavily forested with species characteristic of southern bottomland hardwood communities, including bald cypress, nuttal, and cherry bark oak. Impacts of river floodplain development include forest loss and water gain in northern reaches and grassland and forest losses in the south.

European settlement in the upper Midwest region brought many changes to the landscape and waterways. The rivers provided efficient transportation and were the focal point of commerce and colonization. As the Midwest economy and population grew, so did the demand for water transport. The US government became involved in Mississippi River navigation in 1824 when the USACE was tasked with removing logs and other obstructions from the river channels to ease constraints on steamboat travel, which was very hazardous.

13.3.2 Description

The UMR extends from the confluence with the Ohio River, River Mile 0.0, to Upper St. Anthony Falls Lock in Minneapolis–St. Paul, Minnesota, River Mile 854.0. The Illinois Waterway extends from its confluence with the Mississippi River at Grafton, Illinois, River Mile 0.0, to T.J. O'Brien Lock in Chicago, Illinois, River Mile 327.0. The UMR-IWW Navigation System contains 1,200 mi of 9 ft deep channels, 37 lock and dam sites, and thousands of channel training structures. The width of the 9 ft channel is generally maintained at 300 ft but may extend to 500 ft on river bends.

The Illinois Waterway is a major tributary of the UMR. It provides navigation from Lake Michigan and Chicago to the UMR, linking the Great Lakes with the inland waterway system.

The UMRS ecosystem consists of hundreds of thousands of acres of bottomland forest, islands, backwaters, side channels and wetland—all of which support more than 300 species of birds, 57 species of mammals, 45 species of amphibians and reptiles, 150 species of fish, and nearly 50 species of mussels. More than 40% of North America's migratory waterfowl and shorebirds depend on the food resources and other life requisites (e.g., shelter, nesting habitats) that the system provides. It is a migratory flyway for 40% of all North American waterfowl.

13.3.3 Cumulative Effects

The analysis and understanding of cumulative effects acting on the UMRS ecosystem presented an important context for developing the ecosystem restoration alternatives. The historic change in land cover (habitat) diversity, resulting from cumulative effects, informed the creation of a virtual reference for ecosystem sustainability. The identification and qualification of habitat-altering processes that will continue to affect the system in the future helped establish both the level and type of measures needed for ecosystem maintenance and restoration.

The without-project future for the UMRS ecosystem would include fewer backwater acres; less water depth in nonchannel habitats, degraded forest structure, and land cover diversity; and uncoordinated floodplain management.

The natural resource managers identified deep backwaters, grasslands, hardwood forests, and marsh habitats as most threatened. River regulation, sedimentation, and floodplain development were rated as the primary stressors. The public identified water quality, sedimentation, and backwater and wetland degradation as significant problems. The game and nongame animals that depend on the diverse river ecosystem would decline commensurate with the decline of river habitats.

There has been a gradual decline in the UMRS ecosystem health and quality. Current levels of environmental management and restoration have not prevented systemwide habitat degradation in the past and will likely not meet existing habitat needs in the future. Increased efforts to reverse impoundment effects on aquatic habitats, vegetation succession, and forest health will be required to sustain ecosystem values. (USACE 2004).

13.3.4 Restoration Plan

The ecosystem restoration plan was authorized in Title VIII of Water Resources Development Act 2007. The plan included actions and acquisitions to address the cumulative and ongoing effects of the navigation system.

The authorized ecosystem restoration first increment plan includes about 225 projects in three categories:

1. Fish Passage and Dam Point Control. The authorized plan includes construction of fish passage at Dams 4, 8, 22, and 26 on the UMR along with engineering and design for fish passage at Dam 19 on the UMR. Dam point control will augment hinge point control for water-level management and is authorized at Dams 16 and 25 on the UMR. The total authorized cost for fish passage is $245 million, and $48 million is authorized for dam point control.
2. Ecosystem Restoration Projects Below Ordinary High-Water Mark or in connected backwater that modify the operation of structures for navigation or that are located on federally owned land. This consists of about 210 projects as generally described in the feasibility report that are located below ordinary high water or on connected backwater, that modify the operation of structures for navigation, or that are located on federally owned land. These projects represent approximately $1.1 billion of the total initial authorization of $1.7 billion for ecosystem restoration and include water-level management, island building, backwater restoration, side channel restoration, wing dam alteration, island and shoreline protection, topographic diversity improvement, and dam embankment lowering.
3. Ecosystem Restoration Projects Involving Land and Easement Acquisition, primarily floodplain restoration projects. This consists of about 35,000 ac of floodplain acquisition for purposes of floodplain connectivity and wetland and riparian habitat protection and restoration at an estimated total cost of $300 million (USACE 2008).

13.3.5 Adaptive Management

The UMR-IWW project will be implemented under an incremental AM approach. The AM approach will focus on delivering meaningful navigation and restoration benefits as early as possible, scheduling projects to provide early benefits and learning that can be applied to future projects, scheduling projects recognizing their mutual dependency in realizing navigation and ecosystem restoration system benefits, and phasing large projects to provide early benefits (USACE 2008).

13.4 LOCKS 27 MISSISSIPPI RIVER MAJOR REHABILITATION

Locks 27 are located in the Chain of Rocks canal adjacent to the Mississippi River near St. Louis, Missouri. The lock site consists of a main chamber

Figure 13-1. Lock 27.
Source: USACE, St. Louis District.

1,200 ft long × 110 ft wide and an auxiliary chamber 600 ft long × 110 ft wide. The locks were built between 1946 and 1953. Figure 13-1 shows Locks 27.

The dam was built several miles away on the main channel of the Mississippi River. The dam is fixed crest (no gates) and was built between 1959 and 1964.

Over the years, the aging of the lock structure required a major rehabilitation effort to increase its functional life. The 5-year rehabilitation project was completed in 2013 at a cost of $52.9 million. The EA was completed in January 2002. The following features were rehabilitated:

- Main lock gate: Install a new steel gate,
- Main lock lift gate downstream leaf: The gate is 30 ft high × 110 ft wide and weighs 225 tons,
- Culvert valves: Replace both fill and empty valves for both locks at $3.5 million,
- Lock wall stability: A steel reinforcing bar would anchor the concrete lock walls to the underlying rock substitute,
- Upstream protection cells: New cement-filled cells would replace existing rock-filled cells,
- Downstream sill stability anchorage: New anchors with corrosion protecting would be installed and existing anchors abandoned,
- Culvert valve machinery: Replace tainter valve machinery at both the main lock and the auxiliary lock,

- Liftgate machinery: Main and auxiliary locks have lift gate on the downstream end of the lock chamber. Each gate has two leafs. This item would replace machinery for both of the downstream leafs,
- Lower canal entrance: Remove existing rock dike and replace with a series of bendway weirs, and
- Lock bulkheads: Replace 14 bulkheads used as temporary structures for dewatering locks.

Environmental categories and environmental impact of the proposed project include

- Aquatic,
- Terrestrial,
- Air quality,
- Noise,
- Hazardous and toxic material disposal,
- Socioeconomic considerations,
- Prime farmland,
- Recreational traffic,
- Navigation industry, and
- Increased navigation capacity.

Each of these subjects is evaluated to identify adverse or no impact caused by the project. The next part of the EA is to identify all listed endangered species in the vicinity of Lock 27 and the impact of the project on each species.

The US Fish and Wildlife Service provided a list of endangered species with classification and habitat:

Classification	*Common and Scientific Name*	*Habitat*
Endangered	Gray bat (*Myotis grisescens*)	Caves
Endangered	Indiana bat (*Myotis sodalis*)	Caves, mines, small stream corridors with well-developed riparian woods; upland forests
Threatened	Bald eagle (*Haliaeetus leucoccephalus*)	Breeding and wintering along major rivers and reservoirs
Endangered	Least tern (*Sterna antillarum*)	Bare alluvial and dredged spoil islands and sand/gravel bars

(Continued)

Classification	*Common and Scientific Name*	*Habitat*
Endangered	Pallid sturgeon (*Scaehirhynchus albus*)	Large rivers
Threatened	Decurrent false aster (*Bolitonia decurrens*)	Disturbed alluvial soils

The St. Louis District opinion was that the proposed project will not adversely impact any of the threatened species in the Lock 27 project area. This EA was given to the USFWS for review and comment.

The relationship of the plan to environment requirement was included in the EA as follows:

Guidance	*Degree of Compliance*
Clean Air Act, 42 USC 7401-7542	FC
Clean Water Act, 33 USC 1251-1375	FC
Comprehensive Environmental Response, Compensation, and Liability Act, 42 USC 9601-9675	FC
Endangered Species Act, 16 USC 1531-1543	PC[1]
Farmland Protection Policy Act, 7 USC 4201-4208	FC
Fish and Wildlife Coordination Act, 16 USC 661-666c	PC[1]
Floodplain Management, EO 11988 as amended by EO 12148	FC
Food Security Act of 1985, 7 USC varies	FC
Land and Water Conservation Fund Act, 16 USC 460d-4601	FC
National Environmental Policy Act, 42 USC 4321-4347	PC[1]
National Historic Preservation Act, 16 USC 470 *et seq.*	FC
Noise Pollution and Abatement Act, 42 USC 7691-7642	FC
Prevention, Control, and Abatement of Air and Water Pollution at Federal Facilities, EO 11282 as amended by EO's 11288 and 11507	FC
Protection and Enhancement of Environmental Quality, EO 11991	FC
Protection and Enhancement of the Cultural Environment, EO 11593	FC
Protection of Wetlands, EO 11990 as amended by EO 12608	FC
Resource, Conservation, and Rehabilitation Act, 42 USC 6901-6987	FC

(Continued)

Guidance	*Degree of Compliance*
Rivers and Harbors Appropriation Act, 33 USC 401-413	FC
Water Resources Development Acts of 1986, 1990, and 1992	FC

Note: FC = Full Compliance, PC = Partial Compliance
[1]Full compliance will be attained upon completion of any permitting requirements or coordination with other agencies.

The EA was sent to the following agencies, elected officials, organizations, and individuals for review and comment:

- USFWS,
- EPA,
- Missouri Department of Natural Resources,
- Mayor, Granite City,
- Illinois Historic Preservation Agency,
- Illinois Environmental Protection Agency,
- Missouri Department of Conservation,
- Illinois Department of Natural Resources,
- Sierra Club,
- The Nature Conservancy, and
- Tri-City Regional Port District. (USACE 2002).

13.5 BONNEVILLE SECOND POWERHOUSE

Bonneville Dam is located on the Columbia River 40 mi east of Portland, Oregon. The first powerhouse, spillway, and original lock were completed in 1938. A second powerhouse was completed in 1981 and a larger navigation lock in 1993.

When Congress approved the construction of a second powerhouse at Bonneville, the Corps asked the NPS to conduct surveys to identify any valuable archaeological sites in the affected area. In 1974 archaeologists from the University of Washington, who were working under contract with the NPS, located five sites that they considered archaeologically significant. Four of those sites could be avoided by a change of construction plans, but one site on the Columbia River at North Bonneville would be covered by water. In 1976 university researchers organized an archaeological testing program at the site that was to be flooded. Based on the results of the testing program, the COE, along with the Advisory Council on Historic Preservation, the Washington State Historic Preservation

Officer, and the NPS, agreed to sponsor large-scale archaeological investigations to recover scientific information contained in the site before it was destroyed.

To begin the process of investigation, the COE hired Environment Consultants, Inc., from Dallas, Texas, to excavate the site. With a crew of 30 excavators, the contractors worked from December 1977 through May 1979, unearthing more than 500,000 artifacts. More than 600,000 artifacts recovered at the site documented the remains of a major aboriginal village once visited by explorers Lewis and Clark in 1805–1806. Archaeologists categorized the artifacts into groups, including stone, wood, bone, metal, glass, floral and faunal remains, and perishable items (leather, cloth, and wood). They also discovered remnants of two types of Indian homes—pit houses and plank houses. William Clark had described these structures in his journal (Clark 1805). "Usually a pit was dug one to four feet deep," he wrote, "the wall planks set vertically to the eaves, a small hole left in one end for a door, and an opening in the roof for the smoke to escape—several families occupied one house."

Researchers believed that the house pits located at the North Bonneville site were the same ones that the Lewis and Clark Expedition observed on their trip down the Columbia River. "I passed four large houses on the Star side a little above the last rapid and opposite a large island which is situated near the Lar Side," wrote William Clark in October 1805.

A large number of artifacts preserved at the site were unique. The site at North Bonneville was also special because of its size and location. Situated on the Columbia River, the site was a major fishing village and a critical link on the Columbia River trade route. This is the only known undisturbed site on the lower Columbia that contained evidence of occupation from prehistoric to recent historic times.

The initial step in the curation is to clean, stabilize, and package the artifact. Then the artifact is described and cataloged. Following the initial curation state, the COE had to decide where the artifacts would be housed. The agency usually made arrangements with public and private institutions, such as museums and universities, to store relics obtained on COE sites. In the case of the North Bonneville site, which revealed an enormous quantity of artifacts, the COE determined that no adequate facilities were available. The Bonneville Auditorium, located on the grounds of the Bonneville Lock and Dam Project, was selected to house the artifacts. The COE opened the Curation Center on April 24, 1989.

The artifacts, however, did not remain at the center. In the 1990s, the COE turned them over to the Yakama Nation in south-central Washington as a result of a cooperative agreement between the District and the Nation. The agreement called for the Yakamas to curate the artifacts, with the COE continuing to pay for any general management costs. The Yakama Nation continued to allow researchers to access the artifacts for their work (USACE 2003b). Figure 13-2 shows the Bonneville Lock and Dam complex. The second powerhouse is in the foreground, spillway in the center, and the first powerhouse and lock are in the background.

Figure 13-2. Bonneville project.
Source: USACE, Portland District.

13.6 COLUMBIA RIVER CHANNEL IMPROVEMENT PROJECT

13.6.1 Project

The CRCIP was to deepen a 40 ft deep ship channel to 43 ft. The 110 mi channel extends from the Pacific Ocean at Astoria, Oregon, to the inland port of Portland, Oregon. The channel deepening was only on the segments that were less than 43 ft deep. The majority of the channel is naturally 50 to 100 ft deep. The project dredged more than 15 million yd^3 of material and cost $182 million. The timeline for this project covered between 1999 and 2010 (USACE n.d.):

1999	Feasibility Report and EIS
2003	Supplemental EIS
2005	Start construction
2006	Start development of AM plan
2010	Complete construction

13.6.2 Original Environmental Impact Statement and Supplemental Environmental Impact Statement

The 1999 EIS found no significant impacts on fish or wildlife. In December 1999, NOAA Fisheries (National Marine Fisheries Service) issued a No Jeopardy Biological Opinion on the expected impacts to salmonids, and the USFWS completed its No Jeopardy Biological Opinion on the potential impacts to wildlife and plant species. In August 2000, NOAA Fisheries withdrew their opinion citing the availability of new information regarding impacts to bathymetry (water depths) and flow on estuarine habitat and resuspension of contaminants. However, the USFWS Biological Opinion remains valid. Because a Biological Opinion that meets the ESA requirements for listed salmonids must be in place before the project can proceed, the Corps and NOAA Fisheries began a consultation process to resolve the issues; the USFWS also reentered the process for two aquatic species—coastal cutthroat trout and bull trout.

In January 2002, the COE submitted the BA to the NOAA Fisheries and USFWS. The 2001 BA included actions associated with dredging and deepening, including compliance measures to minimize incidental take of listed species, monitoring actions to ensure that deepening and disposal have minimal effects on listed fish and their habitats, and AM to respond to impacts discovered through the monitoring program. The BA also included ecosystem restoration features and evaluation actions involving numerous proposals to improve existing habitat conditions in the lower Columbia River and estuary and evaluation activities to increase knowledge of the river and estuary ecosystem.

Several other steps remain before project construction would begin. The WDOE and the Oregon Department of Environmental Quality must issue Section 401 Water Quality certifications under the CWA, and the WDOE and Oregon Department of Land Conservation and Development must evaluate the proposed action for consistency under the CZMA. Both states initially denied Section 401 certification and CZMA consistency in 2000. Since then, the COE and Sponsor Ports have met repeatedly with officials from Washington and Oregon to understand and work to address the issues identified by the agencies. The COE has applied for 401 Certification and has submitted CZMA Consistency Determinations. Coordination between the COE and these state agencies is ongoing.

This Final SEIS also includes an updated benefit–cost analysis for the project. The updated analysis was conducted between January and June 2002 and focuses on confirming what are the benefits and costs of the 43 ft channel. Each of the inputs to the benefit and cost calculations were reviewed and updated using the most current data available.

The Final SEIS focused on

- Navigation channel improvements: The Final SEIS reflects the decision to defer action on deepening the Willamette River until after EPA decisions have been made regarding the cleanup of the parts of the river listed as a Superfund site. The Final SEIS, therefore, focuses on the Columbia River; impacts regarding the Willamette River are discussed to a lesser extent in Section

6.12. With regards to new information, much of the new information presented in the Final SEIS pertains to impacts of deepening the Columbia River, called the Channel Improvements Project.

- Restoration projects: The Final SEIS reflects the incorporation of five new restoration features and analyzes the environmental impacts associated with implementing these features. The new restoration features result in a minor change to long-term disposal needs.
- Long-term disposal needs for MCR and Channel Improvements projects: The Final SEIS discusses revisions to upland disposal sites for the Channel Improvements Project that resulted from the consultation process with NOAA Fisheries. In addition, implementation of the proposed restoration features at the Lois Mott embayment and Millar Pillar are anticipated to significantly reduce the need for ocean disposal of river channel material.

On May 20, 2002, NOAA Fisheries and the USFWS transmitted their final Biological Opinions to the COE. These opinions determined that the Channel Improvement Project, including dredging, disposal, monitoring, AM, evaluation, and ecosystem restoration, is not likely to jeopardize the continued existence of 13 listed and 1 proposed fish species, bald eagles, or Columbian white-tailed deer. The additional project features or actions would not affect other species addressed in the 1999 BA for the Channel Improvement Project. In addition, the NOAA Fisheries concurred that the project is not likely to adversely affect Steller sea lions (USACE 2003a).

13.6.3 Adaptive Environmental Management

The AEM program continued into the first few years of the O & M phase to evaluate postconstruction monitoring results. The following AMT consensus decisions regarding the continued implementation of the CRCIP AM program resulted from AMT discussions that continued throughout 2013 and subsequently to the conclusion of the AEM program in 2014 (USACE 2015):

- MA-1: This concluded that changes to physical parameters as a result of the project are minor and occur in proximity to the navigation channel.
- MA-2: This concluded that dredging volumes did not exceed planned placement capacity.
- MA-3: This concluded that river bottom side-slope adjustments caused by dredging occurred as expected (intermittently adjacent to the navigation channel).
- MA-4: This concluded that modeling results did not indicate a significant change to shallow water habitat or salmon habitat opportunity.
- MA-5: This concluded that dredging would not expose aquatic organisms to toxic contaminants.
- MA-6: This concluded that modeling results did not indicate an increase in fish stranding.

- Bank-to-Bank Survey: The postconstruction survey 2 years after construction required by WDOE was completed in 2012.
- Sediment Management: This program component addressed concerns about potential impacts to valued coastal zone habitats caused by placement of dredged material. Nearshore sites were given priority over estuarine and deep-water ocean placement. The USACE regional sediment management program will continue into the operations and maintenance phase.
- Sturgeon: This program component addressed concerns about potential impacts to sturgeon. Green sturgeon tagging and detection were performed in 2010.
- Crab: This program component addressed concerns about potential impacts to Dungeness crab. No change was determined since 2011.
- Smelt: This program component addressed concerns about potential impacts to smelt (eulachon). No change in smelt was found.

13.7 GREAT LAKES RESTORATION

13.7.1 General

The Great Lakes navigation system is a 2,400 mi deep-water network of dredged channels, locks, and open water lake segments. The system extends from Duluth, Minnesota, to the Atlantic Ocean. Over the years, growth of industries and farming has degraded the water quality and reduced habitat necessary for a sustainable ecosystem. To correct past ecosystem damage, a restoration project was launched in 2010. This project has been guided by an interagency task force and a regional working group, which are led by the EPA. These federal agencies include COE, NOAA, USFWS, Coast Guard, Forest Service, and Geologic Survey. About $1.6 billion was spent on restoration projects during the 2010–2014 fiscal years.

13.7.2 Phase I

The first phase of this restoration was 2010 through 2014. The focus areas were

- Toxic substances,
- Invasive species,
- Nonpoint pollution,
- Habitat and wildlife, and
- Education and monitoring.

Toxic substances

Removal of 42 beneficial use impairments in 17 areas of concern—quadrupling the number of beneficial use impairments removed in the preceding 22 years. These beneficial use impairments—benchmarks of environmental harm—include beach closings, restrictions on drinking water consumption, nuisance algal blooms, restrictions on dredging, fish and wildlife deformities, restrictions on fish and wildlife consumption, and loss of fish and wildlife habitat.

Invasive species

During the first 5 years of the Great Lakes Restoration Initiative (GLRI), federal agencies and their partners engaged in an unprecedented level of activity to prevent new introductions of invasive species in the Great Lakes ecosystem. Efforts by agencies and their partners helped prevent bighead and silver carp from becoming established in the Great Lakes ecosystem. Surveillance programs formed the foundation for a multispecies early detection network. Partner agencies responded to several detections, including red swamp crayfish in Wisconsin, grass carp in Michigan, Hydrilla in New York, and silver and bighead carp in the Chicago Area Waterway System.

Federal agencies and their state partners have reduced the risk of invasive species entering the Great Lakes from ballast water discharges. No new introductions have been detected through the ballast water pathway since 2006. Federal agencies and their partners have conducted species risk assessments for organisms posing threats to the Great Lakes ecosystem. Public education efforts have helped boaters, anglers, and other resource users to prevent the spread of invasive species.

The GLRI, federal agencies, and their partners—through their efforts to control and reduce the migration of invasive species—have achieved target levels for controlled populations of invasive species, including

- Baby's breath,
- Bighead carp,
- Buckthorn,
- Emerald ash borer,
- Eurasian watermilfoil,
- Garlic mustard,
- Grass carp,
- Japanese barberry,
- Japanese knotweed,
- Lyme grass,
- Invasive strains of Phragmites,
- Purple loosestrife,
- Silver carp,
- Sea lamprey, and
- Wild parsnip.

No new invasive species have been established since 2009. These control projects were implemented with partners who will continue maintenance and stewardship beyond the duration of the federally funded projects.

Nonpoint pollution

Federal agencies and their partners targeted activities to reduce the largest human-caused nonpoint source of phosphorus inputs to Great Lakes nearshore areas: nutrient runoff from agricultural lands. Excess phosphorus loadings threaten the Great Lakes ecosystem by contributing to harmful algal blooms that can cause human health effects, drinking water impairments, and beach closures; exacerbate dead zones; and result in loss of recreational opportunities. In the summer of 2014, EPA provided almost $12 million to protect public health by targeting harmful algal blooms in western Lake Erie. Federal agencies and their partners provided farmers with financial and technical resources to implement conservation systems to reduce nutrient runoff and control soil erosion. Federal agencies and their partners targeted 720,000 ac of agricultural lands, increasing by more than 70% the number of acres under conservation practices across all three GLRI priority watersheds.

Habitat and wildlife

Federal agencies and their partners worked to protect, restore, and enhance habitat in the Great Lakes basin. Projects were implemented to maintain healthy populations of native species in aquatic and terrestrial habitats.

More than 875 habitat protection, restoration, and enhancement projects were implemented throughout the Great Lakes basin by federal agencies and their partners. More than 100,000 ac of wetlands and 48,000 ac of coastal, upland, and island habitat were protected, restored, and enhanced. More than 500 barriers were removed or bypassed in Great Lakes tributaries, enabling access by fish and other aquatic organisms to more than 3,400 additional miles of river. Data were also collected to document baseline conditions for fish, amphibians, invertebrates, birds, plants, and water quality for all coastal wetlands to inform protection and restoration decisions.

During the first 5 years of the GLRI, federal agencies and their partners worked to maintain, restore, and enhance populations of native fish and wildlife species. To conserve native species that were once broadly distributed across the lakes, the agencies and partners

- Assisted with the delisting of the federally endangered Lake Erie water snake;
- Improved conditions for the following endangered and threatened species: bog turtle, Canada lynx, copperbelly water snake, Eastern Massasauga rattlesnake, Hines emerald dragonfly, Karner blue butterfly, Kirtland's warbler, lakeside daisy, Mitchell's satyr butterfly, piping plover, and Pitcher's thistle; and
- Implemented projects that led to an additional 13 populations of managed native aquatic nonthreatened and nonendangered species becoming self-sustaining in the wild.

Education and monitoring

In response to the administration's goals for improved transparency and fiscal stewardship, federal agencies established accountability mechanisms, management practices, and third-party oversight to effectively manage the GLRI. Section IV includes more information on efforts to ensure accountability.

The GLRI funding continues to enhance existing programs that assess the physical, biological, and chemical integrity of the Great Lakes. These programs, in coordination with complementary state and Canadian programs, help to evaluate the effectiveness of restoration efforts and to assess the overall health of the Great Lakes ecosystem using the best available science. The GLRI has been able to leverage resources and establish a large community of partners to ensure that these efforts are efficient and effective.

During the first 5 years of the GLRI, federal agencies and their partners implemented a number of efforts to promote Great Lakes–based environmental education and stewardship, including the following:

- The Center for Great Lakes Literacy was established by the Great Lakes Sea Grant Network to develop a community of Great Lakes–literate educators, students, scientists, environmental professionals, and citizen volunteers dedicated to improved Great Lakes stewardship.
- The Great Lakes Bay Watershed Education and Training Program was created to promote hands-on environmental activities that are aligned with academic learning standards.

Collectively, the Center for Great Lakes Literacy, Great Lakes Bay Watershed Education and Training Program, and other education projects have resulted in more than 1,500 educational institutions incorporating Great Lakes–specific material into their broader environmental education curricula. It is estimated that more than 175,000 students have participated in these classes.

13.7.3 Planned Activities

The actions planned for fiscal years 2015–2019 will build on restoration and protection work carried out under the first GLRI Action Plan, with a major focus on

- Cleaning up Great Lakes areas of concern,
- Preventing and controlling invasive species,
- Reducing nutrient runoff that contributes to algal blooms,
- Restoring habitat to protect native species, and
- Supporting Great Lakes resilience, education and AM (how we make better investment decisions over time).

GLRI Phase I incorporates a science-based AM framework that will be used to prioritize ecosystem problems to be targeted with GLRI resources, to select projects to address those problems, and to assess the effectiveness of GLRI projects. Measures of progress have been developed to track all actions implemented under Phase I.

GLRI Phase I commits agencies to develop and incorporate climate resiliency criteria in project selection processes. Agencies will develop standard criteria to ensure climate resiliency of GLRI-funded projects.

GLRI Phase I includes feedback for strengthening the GLRI that was contributed by the Great Lakes Advisory Board, the EPA Science Advisory Board, the US Government Accountability Office, the Congressional Research Service, states, tribes, municipalities, and the public through in-person meetings, webinars, and conference calls (EPA 2015).

13.8 LOWER MISSISSIPPI RIVER RESTORATION

The purpose of the LMR restoration project is to improve the habitat of three endangered species: least tern, pallid sturgeon, and fat pocketbook mussel. Figures 13-4, 13-5, and 13-6 show these species.

The authority for this effort is Section 7(a)(1) of the ESA. This act requires all federal agencies to use their authorities as appropriate to carry out programs for the conservation (recovery) of endangered and threatened species.

The federal project to be ecologically restored is the 953 mi navigation channel between the Gulf of Mexico and the confluence of the Ohio and Mississippi rivers (near St. Louis, Missouri). Although the development of the Mississippi River for year-round navigation and flood protection has provided enormous economic benefit, it has also resulted in the degeneration of river habitat.

Under the project, there have been 774 dikes modified to improve the riverine habitat. The dikes' function is to constrict the river, which creates a deeper channel for navigation. These dikes have affected the habitat downstream from the dike

Figure 13-3. Least tern.
Source: USACE, Mississippi Valley Division.

Figure 13-4. Pallid sturgeon.
Source: USACE, Mississippi Valley Division.

Figure 13-5. Fat pocketbook mussel.
Source: USACE, Mississippi Valley Division.

and have closed some secondary channels. A habitat enhancement technique is to construct notches in the dike to allow water flow for the habitat behind (downstream) from the dike. Notching also opens side channels, which were cut off from access to the main channel. Environmental benefits for dike notching are presented in Chapter 10.

This notching program and other in channel training works that are beneficial for habitat enhancement are ongoing under this conservation plan. The plan includes the AM concept to allow habitat evaluations to be part of the decision process to identify and prioritize future environmental enhancement projects (USACE 2013).

13.9 OREGON CHUB

13.9.1 Introduction

The Oregon chub was listed as endangered in 1993. A recovery plan was published in 1998. Critical habitat was designated on March 10, 2010. The species' status has recently improved, and on April 23, 2010, the USFWS changed the ESA classification of the Oregon chub from endangered to threatened. On February 4, 2014, the USFWS announced a proposal to remove the Oregon chub, and its critical habitat, from the list of endangered and threatened species. On February 18, 2015, the USFWS announced the removal of the Oregon chub and its critical habitat from endangered and threatened species, and the Oregon chub became the first fish ever to be deleted due to recovery.

13.9.2 Historical Status and Current Trends

Oregon chub are endemic to the Willamette River Valley of western Oregon. Although information is scarce, the Oregon chub probably occurred throughout the lower elevations of the Willamette River Valley. Historical records indicate that Oregon chub were found as far downstream as Oregon City and as far upstream as Oakridge. Historical records also report that Oregon chub were collected from the Clackamas River, Molalla River, South Santiam River, North Santiam River, Luckiamute River, Long Tom River, McKenzie River, Mary's River, Coast Fork Willamette River, Middle Fork Willamette River, and the Willamette River from Oregon City to Eugene.

When the species was listed in 1993, there were eight known populations. By 2007, there were 38 known populations, and this met the recovery criteria for downlisting (changing the classification from endangered to threatened). The USFWS downlisted Oregon chub to threatened status in 2010. Currently, there are 50 known populations; 19 of these populations have stable or increasing 7-year abundance trends. The improved status is attributed to successful introduction of Oregon chub into new locations within their historical range and the discovery of new, previously undocumented populations. The populations are found in the Santiam River, Middle Fork Willamette River, Coast Fork Willamette River, McKenzie River, and several tributaries to the Willamette River downstream of the Coast Fork/Middle Fork confluence.

13.9.3 Description and Life History

The Oregon chub is a small minnow with an olive-colored back grading to silver on the sides and white on the belly. Adults are typically less than 9 cm (3.5 in.) in length. Scales are relatively large, with fewer than 40 occurring along the lateral line; scales near the back are outlined with dark pigment. Adults feed in the water column on the tiny larvae of aquatic invertebrates, such as mosquitos and other insects. Spawning occurs from the end of April through early August, when water temperatures are between 16°C and 28°C (60°F and 82°F). Only males larger than 25 mm (1 in.) spawn, and males larger than 35 mm (1.4 in.) defend territories in or near vegetation. Females can lay several hundred eggs.

13.9.4 Habitat

Oregon chub are found in slack water off-channel habitats such as beaver ponds, oxbows, side channels, backwater sloughs, low gradient tributaries, and flooded marshes. These habitats usually have little or no water flow, silty and organic substrate, and aquatic vegetation as cover for hiding and spawning. The average depth of Oregon chub habitats is typically less than 2 m (6 ft), and the summer water temperature typically exceeds 16°C (61°F). Adult Oregon chub seek dense vegetation for cover and frequently travel in the midwater column in beaver channels or along the margins of aquatic plant beds. Larval chub congregate in near-shore areas in the upper layers of the water column in shallow areas. Juvenile Oregon chub venture farther from shore into deeper areas of the water column. In the winter months, Oregon chub can be found buried in the detritus or concealed in aquatic vegetation. Fish of similar size classes school and feed together. In the early spring, Oregon chub are most active in the warmer, shallow areas of the ponds.

13.9.5 Reasons for Decline

Historically, the main stem of the Willamette River was a braided channel with many side channels, meanders, oxbows, and overflow ponds that provided habitat for the chub. Periodic flooding of the river created new habitat and transported the chub into new areas to create new populations. The construction of flood control projects and dams, however, changed the Willamette River significantly and prevented the formation of chub habitat and the natural dispersal of the species. Other factors responsible for the decline of the chub include habitat alteration; the proliferation of non-native fish and amphibians; accidental chemical spills; runoff from herbicide or pesticide application on farms and timberlands or along roadways, railways, and powerline rights-of-way; the application of rotenone to manage sport fisheries; desiccation of habitats and unauthorized water withdrawals; diversions, or fill and removal activities; sedimentation resulting from timber harvesting in the watershed; and possibly the demographic risks that result from a fragmented distribution of small, isolated populations. The introduction of non-native fish and amphibians continues to threaten existing populations of Oregon chub; many non-native species (such as bass, mosquito fish, and bullfrogs) occur in the same type of habitat as Oregon chub and eat small fish, including the chub.

13.9.6 Conservation Measures

In 1998, the USFWS published a recovery plan for the Oregon chub. The goal of this plan is to reverse the decline of the Oregon chub by protecting existing wild populations, reintroducing chub into suitable habitats throughout its historic range, and increasing public awareness and involvement. The US Forest Service, USACE, and the Oregon Department of Fish and Wildlife have active programs to protect the Oregon chub. Careful and coordinated planning, management, and protection of Oregon chub habitat is necessary for the survival of this little minnow (USFWS 2014).

References

Clark, W. 1805. *Journals of Lewis and Clark*. Accessed October 25, 2018. https://lewisandclarkjournals.unl.edu/item/lc.sup.ronda.01.

EPA. 2015. *Great Lakes restoration initiative report to congress and the president, fiscal years 2010–2014*. Washington, DC: EPA.

USACE. 1973. *Final environmental statement red river waterway, Louisiana, Texas, Arkansas, and Oklahoma, and Related Projects*. New Orleans: US Army Engineer District.

USACE. 1983. *Final report and final environmental impact statement for acquisition of wildlife mitigations lands (Red River Waterway)*. New Orleans: US Army Engineer District.

USACE. 2002. *Environmental assessment with draft findings of no significant impacts locks No. 27 major rehabilitation Mississippi River, Granite City, Illinois*. St. Louis, MO: US Army Engineer District.

USACE. 2003a. *Columbia River channel improvement project final supplemental integrated feasibility report and environmental impact statement*. Omaha, NE: US Army Engineer Division.

USACE. 2003b. *Currents of change: A history of the Portland District, US Army Corps of Engineers 1980–2000*. Portland, OR: US Army Engineer District.

USACE. 2004. *UMR-IWW system navigation feasibility study final integrated feasibility report and PEIS*. Washington, DC: USACE.

USACE. 2008. "Implementation guidance for Upper Mississippi River and Illinois waterway system." July 2 memo for the Director of Civil Works from Assistant Secretary of Army (Civil Works). Planning.usace.army.mil/toolbox/library/WRDA/ImpGuide8000.pdf

USACE. 2013. *Conservation plan for interior least tern, pallid sturgeon, and fat pocketbook mussel in the Lower Mississippi River*. Vicksburg, MS: US Army Engineer Division.

USACE. 2015. *Adaptive environmental management for the Columbia River Channel Improvement Project: Annual report for 2013–2014 and final project report*. Portland, OR: US Army Engineer District.

USACE. n.d. *Environmental engineering for shallow draft waterways*. Draft manual. Washington, DC: USACE.

USFWS (US Fish and Wildlife Service). 2014. *Oregon chub*. Accessed October 25, 2018. https://www.fws.gov/oregonfwo/articles.cfm?id=149489414.

APPENDIX A
Acronyms

ACOPNE	Academy of Coastal, Ocean, Ports, and Navigation Engineers
AEM	Adaptive Environment Management
AM	Adaptive Management
AMT	Adaptive Management Team
ATF	Aquatic Transfer Facility, Dredging
BA	Biological Assessment
CAD	Confined Aquatic Disposal, Dredging
CDF	Confined Disposal Facility
CEQ	Council on Environmental Quality
CFR	Code of Federal Regulations
COE	Corps of Engineers (USACE)
COPRI	Coastal, Ocean, Ports and Rivers Institute
CRCIP	Columbia River Channel Improvement Project
CWA	Clean Water Act, short for Federal Water Pollution Act of 1972
CZMA	Coastal Zone Management Act
DDT	Pesticide (dichloro-diphenyl-trichloro-ethane)
DMMP	Dredged Material Management Plan
DMMS	Dredged Material Management Strategy
EA	Environmental Assessment
EIS	Environmental Impact Statement
EM	*Engineer Manual*, USACE
EPA	Environmental Protection Agency
ER	Engineer Regulation, USACE
ESA	Endangered Species Act
FONSI	Finding of No Significant Impacts
GIWW	Gulf Intracoastal Waterway
ITM	Inland Testing Manual
LMR	Lower Mississippi River
LTMS	Long-Term Management Strategy, Dredging
MCR	Mouth of Columbia River
MLLW	Mean Lower Low Water, Tide Datum West Coast United States
MMPA	Marine Mammal Protection Act
MOP	Manual of Practice, ASCE
MPRSA	Marine Protection, Research, and Sanctuaries Act

NAVD	North American Vertical Datum
NE	Navigation Engineering
NED	National Economic Development
NEPA	National Environmental Policy Act
NFH	National Fish Hatchery
NOAA	National Oceanic and Atmospheric Agency
NPS	National Park Service
O & M	Operation and Maintenance
OTM	Ocean Testing Manual
PDF	Probability Density Function
PIANC	World Association for Waterborne Transport Infrastructure, previously Permanent International Association of Navigation Congresses
PL	Public Law
RSM	Regional Sediment Management, Dredging
SEIS	Supplemental Environmental Impact Statement
TSS	Total Suspended Solids, Dredging
UMR	Upper Mississippi River
UMR-IWW	Upper Mississippi River – Illinois Waterway
UMRS	Upper Mississippi River System
USACE	US Army Corps of Engineers
USFWS	US Fish and Wildlife Service
USGS	US Geological Survey
UTM	Upland Testing Manual, Dredging
WDOE	Washington Department of Ecology
WID	Water Injection Dredging
WQMMP	Water Quality Management and Monitoring Plan, Dredging
WRDA	Water Resources Development Act

APPENDIX B

Tennessee–Tombigbee (Tenn–Tom) Waterway Project Evolution

THE PROJECT

The Tenn–Tom Waterway is a 234 mi-long inland waterway providing a navigation connection between the Tennessee River (and thus the Cumberland, Ohio, and Mississippi rivers) and the Gulf of Mexico via the Black Warrior–Tombigbee Waterway and Mobile Bay. It passes through Mississippi and Alabama, as shown in Figure B-1. The USACE began construction in December 1972 and completed the project in December 1984.

The project's congressionally authorized purposes include navigation and wildlife conservation. It also provides extensive recreation and some water supply benefits to the extent that those uses do not adversely affect the authorized purpose benefits. It is not intended to provide flood damage reduction benefits; however, some residents of the Tombigbee River Valley assert that flood damages are noticeably reduced since construction of the waterway.

Since at least 1770 (Ward 2010), individuals had suggested that a shortcut from the Tennessee River to the Gulf of Mexico could be built via a canal to the Tombigbee River basin. Congressional authorizations in 1875 and 1913 led to surveying of a proposed route, but the expected economic benefits did not justify the project. The US Congress authorized the Tenn–Tom Waterway in 1946, but funds were not provided until 1971.

ENVIRONMENTAL SETTING

Prior to waterway construction, the Tombigbee River was reliably navigable only to Columbus, Mississippi, about 300 river miles upstream from Mobile Bay, Alabama. Controlling depths were 1 ft or less during low water seasons and about 4 ft during high water season. The 20 mi reach between Columbus and Aberdeen,

Figure B-1. Tennessee–Tombigbee Waterway.
Source: USACE (2018).

Mississippi, was navigable only during high water, but adverse currents made navigation extremely hazardous. Most navigation upstream of Columbus was limited to recreational vessels and log rafting (Mississippi SPC 1937).

The waterway, constructed between 1972 and 1984, consists of three distinct sections—river, canal, and divide cut—as shown in Figure B-2. The river portion extends upstream from Mile 217, where the waterway connects to the Black Warrior River to Mile 365 near Amory, Mississippi, generally following the course of the Tombigbee River. The canal section starts at Mile 365 and departs from the Tombigbee River course to trend generally northward to Jamie Whitten (Bay Springs) Lock at Mile 412. The divide cut section connects the canal section to the Tennessee River at Pickwick Lake near the Mississippi–Tennessee boundary.

The 149 mi long river section lies within the Tombigbee River floodplain and generally follows the course of the river. A number of river meanders have been cut off, leaving 71 mi of meander loops that are still connected to the waterway. Four lock and dam structures raise the water level 117 ft. The navigation channel has a bottom width of 300 ft and dredged depths of 9 ft or 12 ft, plus 1 ft of

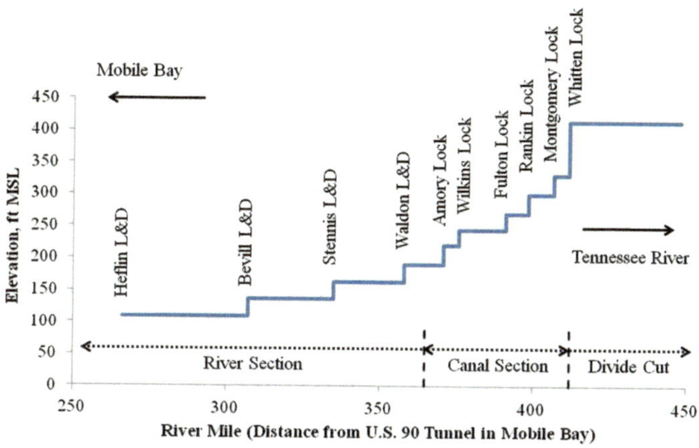

Figure B-2. Tennessee–Tombigbee waterway profile and structures.

allowable overdepth dredging. Numerous tributaries drain into the river section, bringing significant quantities of sediment.

The 46 mi long canal section is located near the eastern edge of the Tombigbee River floodplain and was formed by constructing a levee to serve as the western boundary of the section, whereas natural high ground serves as the eastern boundary. Five pools result in a chain-of-lakes configuration to provide navigable depths with a 300 ft wide by 12 ft deep channel. Inflow to the canal section is limited to discharges from Whitten Lock and small tributaries on the eastern edge of the floodplain.

The divide cut section connects the separate river basins by an excavated cut through the basin divide and extends 39 mi from Bay Springs Lock to Pickwick Lake. The navigation channel has a bottom width of 280 ft and a depth of 12 ft during minimum (winter) pool on Pickwick Lake. Inflows to the section consist of minor local inflows and flow from Pickwick Lake to replace water released downstream at Whitten (Bay Springs) Lock.

Table B-1 lists the pools and structures of the waterway and their dimensions. Each lock or lock and spillway dam forms an upstream pool, which in some cases has the same name as the dam.

Hydrology and Hydraulics

The hydrology of the Tenn–Tom is unusual in that the waters of two basins are mixed by the connection at Bay Springs Lock. There the waters of the Tennessee River, at least the volumes involved in lockages, are conveyed to the previously separate Tombigbee River Valley. This aspect—the mixing of the waters—was one element in a vigorous debate over the waterway's environmental effects prior to and during its construction and the title of a book on the project (Stine 1993).

Table B-2 shows that inflows in the canal section above Pool A are relatively small, averaging only 647 cfs compared with the tributary inflow of 1,397 cfs in Pool A and 2,494 cfs in Aberdeen Pool.

Table B-1. Tennessee–Tombigbee Waterway Navigation Components.

Section	Total length (mi)	Channel width (ft)	Channel depth (ft)	Locks (pool) 110 ft wide × 600 ft long each	Lift (ft)	Normal pool elevation (ft)	Water surface (ac)
River	149	300	9	Howell Lock and Dam (Gainesville)	36	109	6,400
				Tom Bevill Lock and Dam (Aliceville)	27	136	8,300
				John Stennis Lock and Dam (Columbus)	27	163	8,900
				Don Waldon Lock and Dam (Aberdeen)	27	190	4,121
Canal	46	300	12	Amory Lock (Pool A)	30	220	914
				Glover Wilkins Lock (Pool B)	25	245	2,718
				Fulton Lock (Pool C)	25	270	1,642
				Jamie Rankin Lock (Pool D)	30	300	1,992
				G. V. Montgomery Lock (Pool E)	30	330	851
Divide	39	280	12	Whitten Lock (Bay Springs)	84	414	7,645
Total	234				341	—	43,483

Table B-2. Tenn–Tom Average Annual Flows.

Pool	*Upstream inflow*	*Local inflow*	*Discharge outside the waterway*
	301	270	0
Bay Springs	571	70	51
Pool E	590	32	15
Pool D	607	40	0
Pool C	647	447	163
Pool B	931	23	7
Pool A	947	1397	0
Aberdeen	2,744	2,494	0
Columbus	5,238	1,586	0
Aliceville	6,824	689	0
Gainesville	7,315	—	0

Note: Flows are in 1,000 ac-ft.

Annual water flow through the waterway consisting of natural flows plus estimated flow from lockages are shown in Table 2-1.

Sedimentation

Prior to construction of the waterway, the Tombigbee River carried an estimated 2.39 tons of sediment per year at Gainesville, Alabama (Underwood 1985). Table B-3 shows 50% suspended sediment concentration exceedance levels (half the time concentrations were lower, and half the time concentrations were higher) at several measurement stations on the Tombigbee River before construction of the waterway. Much of the watershed experienced severe land erosion and gullying (Mississippi SPC 1937).

Sedimentation issues in the completed waterway were much the same as preconstruction –sheet erosion and gullying during high runoff events. The tributaries of Town Creek, which flows through Tupelo, and Matubby

Table B-3. Tenn–Tom 50% Exceedance Suspended Sediment.

Location	*Concentration (mg/L)*	*Load (tons/day)*
Fulton	129	129
Amory	81	252
Aberdeen	78	258
Columbus	66	400
Aliceville	74	620
Gainesville	37	447

Source: Underwood (1985).

Creek, which flows through Aberdeen, Mississippi, experienced severe channel incising and bank failures, both of which supplied substantial quantities of sediment to the waterway proper. The lower Tombigbee, from Mile 279 seaward, also experienced bankline erosion that Bankhead et al. (2008) examined and could not separate Tenn–Tom Waterway effects from natural causes (e.g., an increase in rainfall over the basin) and prewaterway impoundments (Bankhead et al. 2008).

Total maintenance dredging quantities for the waterway from 1985 through 2001 are given in Table B-4. They show that the average annual dredging quantity for the channel sections of interest was about 825,000 yd^3, with the largest component in Aberdeen Pool at Mile 366, just downstream from the confluence of the waterway and the Tombigbee River (formerly East Fork Tombigbee) channel.

Estimated annual sedimentation rates for each of the public ports within the Mississippi section of the waterway are shown in Table B-5. These estimates represent the average amount of sediment accumulation that might occur if the ports were dredged every year (i.e., maintained to full project dimensions with an annual dredging). The estimates, stated as a rather wide range, were based on the available dredging records by McAnally et al. (2004). Each port is responsible for its own dredging.

Water Quality

The Tombigbee River prior to construction of the waterway was a meandering, shallow stream through forested and farmed lands with overall good to excellent water quality. High suspended sediment concentrations occurred after floods and occasionally low dissolved oxygen (DO) conditions appeared in the summer months. Low alkalinity and moderate nutrient supply kept plankton counts low (Mercante 1980). The areas which became the canal section contained small tributary streams and vegetated lowlands with mostly good water quality. In the

Table B-4. Tenn–Tom Dredged Quantities 1985 through 2001.

Pool	*Dredged volume (yd^3)*
Bay Springs	177,132
E	275,393
D	0
C	0
B	209,216
A	30,652
Aberdeen	3,550,085
Columbus	1,269,829
Aliceville	1,619,807
Total	14,028,705

Source: Compiled from COE unpublished records (1986–2004).

Table B-5. Estimated Range of Annual Tenn–Tom Port Sedimentation.

Port	Low	High	Typical
Yellow Creek	0	0	0
Northeast Mississippi	500	2,500	1,500
Itawamba	3,000	7,000	5,000
Amory	500	3,000	1,500
Aberdeen	5,000	15,000	10,000
Clay County	4,000	12,000	8,000
Lowndes	3,000	10,000	5,000
Total	16,000	49,500	31,000

Note. The sedimentation is in cubic yards.
Source: McAnally et al. (2004).

1970s, some tributary streams, mostly on the west side, suffered poor water quality caused by localized pollution problems from nutrient runoff and sewage outfalls, but they had minor to no impact on the main stem of the river.

Postconstruction water quality continues to be good to excellent in the main stem of the waterway, except for occasional DO readings of about 3 mg/L in the Aliceville pool just upstream of Bevill Lock and dam that fall below Alabama's criterion of 4 mg/L.

Aquatic and Terrestrial Life

The Tombigbee River and its tributaries supported typical warm freshwater riverine and floodplain biota prior to construction, with notable bottomland hardwood forest ecosystems, gamefish, and mussels (IES 1979). Alligators were common in the lower basin and occurred in remote locations in the middle basin. Several threatened and endangered species common to the southeast may have been present, but at the time of construction only five species of mussels were considered by the USFWS to require special consideration (USACE 1983b).

The completed waterway has exchanged modest riverine fisheries for abundant lacustrine fisheries. Mitigation efforts, including maintaining minimum flows in cutoff bends and nearby creeks, were designed to maintain healthy mussel beds and other riverine biota.

Cultural Resources

Significant archeological sites in the near-river basin consist of Native American ceremonial tumuli, shell mounds, and campsites, plus eighteenth- and nineteenth-century plantations, three extinct town sites, and a sunken stern-wheel riverboat. The COE EIS process provided archaeological surveys and curation proposals for all sites of significant interest and provoked little controversy except for the fossil-rich geological formations at Plymouth Bluff near Columbus, Mississippi. In response, the Stennis Lock and Dam was relocated to avoid flooding the site, and the COE constructed a visitor center at the site, which is managed by Mississippi University for Women.

Economics

The economy of the Tombigbee River basin, encompassing sections of western Alabama and northeast Mississippi, was primarily based on agriculture and silviculture, which supported small towns. Ports were limited to intermittently used terminals, such as the Cotton Gin Port above Columbus, with little to no regular commercial traffic and the Port of Gainesville, Alabama. Construction of the waterway led to development of ports in almost every county along the waterway, often with on-site and nearby manufacturing or processing facilities, such as paper mills, a steel mill, and sheet metal fabrication plants.

Economic justification for the project, which at $2 billion was the most expensive public works project in US history at that time, projected net annual benefits of about $137 million, largely based on an expected surge in coal exports. The benefit–cost ratio was estimated in 1971 to be 1.6, but a restudy in 1976 lowered the ratio to 1.08 (Comptroller 1981). At that time, ports and governmental agencies nationwide were gearing up for what seemed to be a burgeoning coal export market.

Tourism in the basin prior to Tenn–Tom construction was minimal, consisting mainly of historical celebrations, antebellum home tours, and visitors to Elvis Presley's birthplace in Tupelo, Mississippi. The completed waterway has seen a boom in recreational uses, including sport fishing, skiing, and pleasure craft excursions, both local and long distance. The project EIS (USACE 1983b) reported a USFWS analysis estimating commercial and recreational fishing revenues of $56,000 per year and projecting a fourfold increase in that number after completion of the waterway. The Columbus, Mississippi, *Dispatch* (2011) reported that a single fishing tournament was expected to bring up to $5 million in economic benefit to that city on the Tenn–Tom. The value of that tournament seems high and was presented by the association hosting the event without supporting documentation, so its accuracy may be questioned. Nevertheless, sport fishing has become a major attraction on the waterway, and the economic benefit should be considerable.

A Troy State University (2009) analysis estimated that from 1996 to 2009, the nation has realized a direct, indirect, and induced economic impact of nearly $43 billion due to the existence and usage of the waterway, a return on initial investment of seven times its construction cost, substantially higher than the COE estimate of less than $3 billion over a 50-year project life.

PLANNING, DESIGN, AND CONSTRUCTION

Studies

Numerous reports document multiple studies to select a final plan, revise the plan, and mitigate adverse environmental impacts of the Tenn–Tom. Water quality, sedimentation, and mussel survey monitoring reports are largely unpublished but retained in the files of federal and state agencies. The COE library at the Engineer Research and Development Center lists 367 documents, and the Special Collections at Mississippi State University contains 938 items that contain the phrase "Tennessee–Tombigbee Waterway" in the title or as a keyword.

These investigations plus controversies over environmental impacts produced a project with an unusual (for the time) dual purpose—navigation and environmental conservation. The Final EIS (USACE 1983b) estimated that 62,800 ac of valuable habitat would suffer adverse impact, including loss of 140 mi of free-flowing river with significant gravel bar habitat, 33,000 acres of bottomland hardwood, 14,000 ac of cropland. The accompanying mitigation plan called for acquisition of 28,000 ac of floodplain with follow-on "intensive management" of the acquired lands plus other federal lands in the basin and creation of minimum flow structures to ensure that the Tombigbee East Fork and other stream segments supported continuing free-flowing stream habitat.

The intensive management plans for habitat in the basin included

- Selective tree cutting,
- Prescribed burns,
- Planting trees for wildlife food sources,
- Creation of open spaces,
- Bird nesting boxes,
- Waterfowl impoundments with water level management for food and cover,
- Beaver pond management,
- Seasonal feeding impoundments,
- Reforestation, and
- Aquatic plant control.

Opposition and Support

The project was opposed by nongovernment organizations such as the Environmental Defense Fund and Louisville and Nashville (now CSX) Railroad, which sued, alleging that the COE had violated the NEPA, exceeded its authority to modify the project during the design phase, and miscalculated the benefit–cost ratio. Federal district and appeals courts ruled against the plaintiffs, and the Final EIS was issued in 1983 (USACE 1983b).

Opposition to the project was primarily based on two issues—potential environmental impacts and cost of the project. Railroad opposition was widely perceived to be based on the threat of competition for freight business but was expressed in terms of environmental concerns.

Environmental concerns raised in the court cases included those previously described in the mitigation plan plus groundwater changes (both drawdown and waterlogging), stratification and eutrophication of the impounded waters, and cross-basin movement of species. The Final EIS (USACE 1983b) addressed those issues but did not placate many project opponents.

Political opposition focused on the project cost and the view that most benefits would accrue to the three-state region of construction. In 1980, US Senators Johnston (LA), Levin (MI), Percy (IL), and Proxmire (WI) and Representatives Edgar (PA) and Pritchard (WA) asked the General Accounting Office to investigate the project, perhaps hoping to accomplish what lawsuits could not—stopping the partially completed project. The US Comptroller General's office (1981) report found some disagreement with the COE's economic analyses but concluded that the COE had correctly followed law and policy. Significantly, the comptroller did not dispute a favorable benefit–cost ratio for the project. The most negative finding was that the project might create a waterway traffic bottleneck on the Black Warrior–Tombigbee River south of Demopolis by 1991 and require another project to permit unrestricted traffic.

Local majorities within Alabama, Mississippi, and Tennessee clearly favored the Tenn–Tom project. When President Carter held a public hearing in Columbus, Mississippi, to gauge support, thousands of citizens from the region attended, offering vociferous support to project completion. Opposition from local citizens included those opposed to land acquisition, by imminent domain if necessary, and conversion of free-flowing stream segments to lakes. The locally formed Committee to Leave the Environment of America Natural joined the Environmental Defense Fund and railroads in opposing the project on multiple grounds, but the opposition could not muster wide support in the region and failed to convince either the courts or Congress to stop the project.

Political support came from the governors, legislatures, and congressional delegations of Kentucky, Tennessee, Mississippi, Alabama, and Florida. Senators Baker (TN) and Stennis (MS) and Congressmen Whitten and Rankin (MS) and Heflin and Bevill (AL) pushed for funding and against efforts to stop the project. Presidents Johnson, Nixon, Ford, Carter, and Reagan each included Tenn–Tom funding in their budget proposals to Congress amid mixed political support from both parties. Each state along the construction route found funds to reroute roads and build new bridges, many of which were built in advance of waterway construction.

LOOKING BACK—PREDICTIONS FULFILLED?

The vantage point of 30 years after Tenn–Tom construction provides some perspective on the much-disputed waterway. First, the project's design and construction straddled the 1970 enactment of the NEPA (PL 91-190) and several related laws, such as the ESA. The COE, along with other agencies and even the courts, were gradually learning the laws' implications and adjusting policies, procedures, and personnel to meet the new requirements. Nine years of litigation informed the processes but delayed construction and increased costs.

The COE's benefits analysis had projected that the Tenn–Tom would carry 28 million tons of cargo, much of it coal, in its first year. However, the 1984 opening was marked by a nationwide economic recession and a decline in coal demand, and the first year's tonnage was only 7 million (*Tuscaloosa News* 2005) and has never reached the expected totals. On the other hand, the value of the cargo tonnage has exceeded estimates because it increasingly includes higher valued cargo such as petroleum products and manufactured goods. The Troy State study (2009) cited benefits (including some indirect benefits the COE could not count under its rules) in just 13 years, totaling 14 times the originally projected 50-year benefits, illustrating the dynamic and often surprising behavior of economies.

Today's waterway, with throngs of recreational boats and sport fishing, is clearly not an environmental disaster as portrayed in public comments by opponents (USACE 1983b). Lowering of groundwater occurred as expected in the divide cut, and a number of well owners were compensated for their losses (McClure 1985). Waterlogging of soils adjacent to the waterway has been minimal (McClure 1985) and caused some changes in flora that are generally considered neutral or beneficial, especially as the value of wetlands has become better understood. Water quality has remained about the same as preconstruction—still good to excellent except for some tributaries polluted by local sources (MDEQ 2001) and with localized low DO during summer months.

Public support for the waterway remains strong locally and is considered a primary asset by regional and state economic development groups, including the Tennessee–Tombigbee Waterway Development Authority, an interstate compact consisting of the states of Alabama, Kentucky, Mississippi, and Tennessee. Political support remains largely dominated by regional interests, with strong support in the southeast and disinterest in the rest of the country. Public opinion outside the southeast, largely ignorant of facts on the ground, still includes a perception of pork barrel politics. The COE has difficulty providing funds for maintenance from a declining budget stretched by nationwide needs for infrastructure maintenance.

References

Bankhead, N., A. Simon, and D. Klimetz. 2008. "Analysis of streambank erosion along the Lower Tombigbee River, Alabama." *Research Rep. No. 62*. Oxford, MS: USDA-ARS National Sedimentation Laboratory.

Comptroller. 1981. *To continue or halt the Tenn–Tom Waterway? Information to help the congress resolve the controversy*. Comptroller General's Report to the Chairman Subcommittee on Energy and Water Development, Committee on Appropriations, US Senate. Gaithersburg, MD: US General Accounting Office.

IES (Institute for Environmental Studies). 1979. *An ecological study of the Tennessee-Tombigbee Waterway*. Report Submitted to the Mobile District, Corps of Engineers. Starkville, MS: Mississippi State Univ.

McAnally, W. H., J. F. Haydel, and G. Savant. 2004. *Port sedimentation solutions for the Tennessee–Tombigbee Waterway in Mississippi*. Starkville, MS: Mississippi State Univ.

McClure, N. D. 1985. "A summary of environmental issues and findings: Tennessee–Tombigbee Waterway." *Environ. Geol. Water Sci.* **7** (1/2): 109–124.

MDEQ (Mississippi Department of Environmental Quality). 2001. *Basin Group 1: Tombigbee River Basin data collection plan 1999–2000*. Jackson, MS: MDEQ.

Mercante, D. E. 1980. "Water quality and plankton study of the upper Tombigbee River along the proposed route of the Tennessee–Tombigbee Waterway." Ph.D. dissertation, Mississippi State Univ., Dept. of Wildlife and Fisheries.

Mississippi SPC (State Planning Commission). 1937. *Report on the Tombigbee River drainage basin*. Jackson, MS: Mississippi State Planning Commission and Works Progress Administration.

Stine, J. K. 1993. *Mixing the waters: Environment, politics, and the building of the Tennessee–Tombigbee Waterway*. Akron, OH: Univ. of Akron Press.

Troy State University. 2009. *Analysis of the economic impact of the Tennessee-Tombigbee Waterway: 1996–2008*. Troy, AL: Troy State Univ.

Tuscaloosa News. 2005. *20-year anniversary of Tennessee-Tombigbee Waterway*. January 9, Tuscaloosa, AL. Accessed October 19, 2018. https://www.tuscaloosanews.com/news/20050109/20-year-anniversary-of-tennessee-tombigbee-waterway.

Underwood, K. D., and F. D. Imsand. 1985. "Hydrology, hydraulic, and sediment considerations of the Tennessee–Tombigbee waterway." *Environ. Geol. Water Sci.* **7** (1–2): 69–90.

USACE. 1983b. *Final report: Wildlife mitigation feasibility study and environmental impact statement for the Tennessee–Tombigbee Waterway, Alabama–Mississippi*. Mobile, AL: US Army Engineer District.

USACE. 2018. *Untitled web page*. Accessed October 19, 2018. https://media.defense.gov/2014/Jul/23/2000807662/-1/-1/0/100126-A-CE999-001.JPG.

Ward, R. 2010. *Tombigbee River*. Palo Alto, CA: History Press.

APPENDIX C

Habitat Development Using Dredged Material

GENERAL CONSIDERATIONS FOR HABITAT DEVELOPMENT

1. Habitat development refers to the establishment of relatively permanent and biologically productive plant and animal habitats. The use of dredged materials as a substrate for habitat development offers a disposal technique that is, in many situations, a feasible alternative to more conventional open-water, wetland, or upland disposal options.
2. Four general habitats are suitable for establishment on dredged material: marsh, upland, island, and aquatic. Within any habitat, several distinct biological communities may occur (Figure C-1). The determination of the feasibility of habitat development will center on the nature of the surrounding biological communities; the nature of the dredged material; and the site selection, engineering design, cost of alternatives, environmental impacts, and public approval. If habitat development is the selected alternative, a decision regarding the type or types of habitats to be developed must be made. This decision will be largely judgmental, but in general, site peculiarities will not present more than one or two logical options.
3. The selection of habitat development as a disposal alternative will be competitive with other disposal options when the following conditions exist:
   - Public/agency opinion strongly opposes other alternatives.
   - Recognized habitat needs exist.
   - Enhancement measures on existing disposal sites are identified.
   - Feasibility has been demonstrated locally.
   - Stability of dredged material deposits is desired.
   - Habitat development is economically feasible.
4. Disposal alternatives are often severely limited and constrained by public opinion and/or agency regulations. Constraints on open-water disposal and

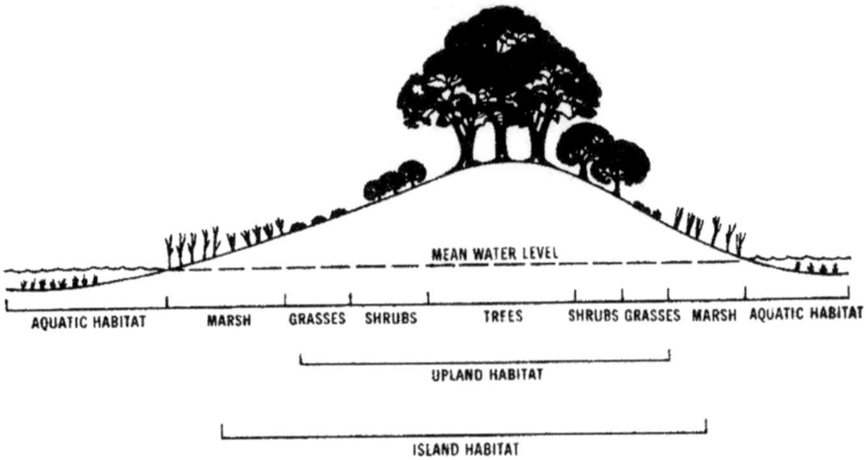

Figure C-1. Hypothetical site illustrating the diversity of habitat types that may be developed at a dredged material placement site.
Source: US Army Corps of Engineers (2015).

disposal on wetlands, or the unavailability of upland disposal sites, may leave habitat development as the most attractive alternative.

5. Habitat development may have strong public appeal when the need for restoration or mitigation or the need for additional habitat has been demonstrated. This is particularly true in areas where similar habitat of considerable value or public concern has been lost through natural processes or construction activities.
6. Habitat development may be used as an enhancement measure to improve the acceptance of a disposal technique. For example, sea grass may be planted on submerged dredged material or wildlife food plants established on upland confined disposal sites. This alternative has considerable potential as a low-cost mitigation procedure and may be used to offset environmental impacts incurred in disposal.
7. The concept of habitat development is more apt to be viewed as feasible if it has been successfully demonstrated locally. Even the existence of a pilot-scale project in a given locale will offset the uncertainties often present in the public perception of an experimental or unproven technique.
8. The vegetation cover provided by most habitat alternatives will often stabilize dredged material and prevent its return to the waterway. In many instances, this aspect will reduce the amount of future maintenance dredging necessary at a given site and result in a positive environmental and economic impact.
9. The economic feasibility of habitat development should be considered in the context of long-term benefits. Biologically productive habitats have

varied but unquestionable value (e.g., sport and commercial fisheries) and are relatively permanent features. Consequently, habitat development may be considered a disposal option with long-term economic benefits that can be applied against any additional costs incurred in its implementation. Most other disposal options lack this benefit.

10. Habitat development may be most economically competitive in situations where it is possible to take advantage of natural conditions or where minor modifications to existing methods would produce desirable biological communities. For example, the existence of a low-energy, shallow-water site adjacent to an area to be dredged may provide an ideal marsh development site and require almost no expenditure beyond that associated with open-water disposal.

Marsh Habitat Development

1. Marshes are considered to be any community of grasses and/or herbs that experience periodic or permanent inundation. Typically, these are intertidal fresh, brackish, or salt marshes or relatively permanently inundated freshwater marshes. Marshes are often recognized as extremely valuable natural systems and are accorded importance in food and detrital production, fish and wildlife cover, nutrient cycling, erosion control, floodwater retention, groundwater recharge, and aesthetic value. Marsh values are highly site specific and must be interpreted in terms of such variables as plant species composition, wildlife use, location, and size, which in turn influence their impact upon a given ecosystem.
2. Marsh creation has been the most studied of the habitat development alternatives, and accurate techniques have been developed to estimate costs and to design, construct, and maintain these systems. More than 100 marshes have been established on dredged material. Refer to WES TR DS-78-16 for specific information on wetland habitat development. The advantages most frequently identified with marsh development are considerable public appeal, creation of desirable biological communities, considerable potential for enhancement or mitigation, and the fact that it is frequently a low-cost option.
3. Marsh development is a disposal alternative that can generate strong public appeal and has the potential for gaining wide acceptance when other techniques cannot. The habitat created has biological values that are readily identified and are accepted by many in the academic, governmental, and private sectors. However, application requires an understanding of local needs and perceptions and of the effective limits of the value of these ecosystems.
4. The potential of this alternative to replace or improve marsh habitats lost through dredged material disposal or other activities is frequently overlooked. Techniques are sufficiently advanced to design and construct

productive systems with a high degree of confidence. In addition, these habitats can often be developed with very little increase in cost above normal project operation, a fact demonstrated by hundreds of marshes that have been inadvertently established on dredged material.

5. The following problems are most likely to be encountered in the implementation of this alternative: unavailability of appropriate sites, loss of other habitats, release of contaminants, and loss of the site for subsequent disposal.
6. The most difficult aspect of marsh development is the location of suitable sites. Low-energy, shallow-water sites are most attractive; however, cost factors will become significant if long transport distances are necessary to reach those sites. Protective structures may be required if low-energy sites cannot be located, which can add considerably to project cost.
7. Marsh development frequently means the replacement of one desirable habitat with another, and this will likely be the source of most opposition to this alternative. There are few reliable methods of comparing the various losses and gains associated with this habitat conversion; consequently, relative impact may best be determined on the basis of the professional opinion of local authorities.
8. The potential for plants to take up contaminants and then release them into the ecosystem through consumption by animals or decomposition of plant material should be recognized when contaminated sediments are used for habitat development. Although this process has not been shown to occur often, techniques are available to determine the probability of uptake.
9. Development of a marsh at a given site can prevent the subsequent use of that area as a disposal site. In many instances, any further development on that site would be prevented by state and federal regulations. Exceptions may occur in areas of severe erosion or where the initial disposal created a low marsh and subsequent disposal would create a higher marsh.
10. There are types of wetland habitat development other than marshes, such as bottomland hardwoods in freshwater areas. These are addressed in WES TR DS-78-16.

Upland Habitat Development

1. Upland habitats encompass a variety of terrestrial communities ranging from bare soil to dense forest. In its broadest interpretation, habitat occurs on all but the most disturbed upland disposal sites. For example, a gravelly and bare freshwater disposal area may provide nest sites for killdeer, weedy growth may provide cover for raccoons or a food source for seed-eating birds, and water collection in desiccation cracks may provide breeding habitat for mosquitoes. Man-made habitats will develop regardless of their management; however, the application of sound management techniques

will greatly improve the quality of those habitats and the speed with which they are populated.

2. Upland habitat development has potential at hundreds of disposal sites throughout the United States. Its implementation is largely a matter of the application of well-established agricultural and wildlife management techniques. Refer to WES TR DS-78-17 for more detailed information on upland habitat development. Upland habitat development as a disposal option has several distinct advantages, including adaptability, improved public acceptance, creation of biologically desirable habitats, elimination of problem areas, low-cost enhancement or mitigation, and compatibility with subsequent disposal.
3. Upland habitat development may be used as an enhancement or mitigation measure at new or existing disposal sites. Regardless of the condition or location of a disposal area, considerable potential exists to convert it into a more productive habitat. For example, small sites in densely populated areas may be keyed to small animals adapted to urban life, such as seed-eating birds and squirrels. Large tracts may be managed for a variety of wildlife, including waterfowl, game mammals, and rare or endangered species.
4. The knowledge that a site will ultimately be developed into a useful area, be it a residential area, park, or wildlife habitat, improves public acceptance. Many idle and undeveloped disposal areas that are now sources of local irritation or neglect would directly benefit from upland habitat development, and such development may well result in more ready acceptance of future disposal projects.
5. In general, upland habitat development will add little to the cost of disposal operations. Standard procedures may involve liming, fertilization, seeding, and mowing. A typical level of effort is similar to that applied for erosion control at most construction sites and considerably less than that required for levee maintenance.
6. Unless the target habitat is a long-term goal such as a forest, upland habitat development will generally be compatible with subsequent disposal operations. In most situations, a desirable vegetative cover can be produced in one growing season. Subsequent disposal would simply require recovery of the lost habitat. Indeed, the maintenance of a particular vegetation state may require periodic disposal to retard or set back plant succession.
7. The primary disadvantage of this alternative is related to public acceptance. The development of a biologically productive area at a given site may discourage subsequent disposal or modification of land use at that site. This problem can be avoided by the clear identification or establishing of future plans before habitat development, or by the establishment and maintenance of biological committees, recognized as being most productive in the earlier states of succession. In the latter case, subsequent disposal may be a necessary management tool.

8. Some habitat types will require management. For example, if high-productivity annual plants are selected for establishment (i.e., corn or barley as prime wildlife foods), then yearly planting will be necessary. If the intent is to maintain a grassland or open-field habitat, planting may be required only initially, but it may be necessary to mow the area every 1 to 5 years to retard colonizing woody vegetation. In most cases, it will be possible to establish very low-maintenance habitats, but if the intent is to establish and perpetuate a given habitat type, long-term management may be essential and expensive.

ISLAND HABITAT DEVELOPMENT

1. Dredged material islands range in size from 1 ac to several hundred acres. Island habitats are terrestrial communities surrounded by water or wetlands and are distinguished by their isolation and their limited food and cover. Because they are isolated and relatively predator free, they have particular value as nesting and roosting sites for numerous species of sea and wading birds (e.g., gulls, terns, egrets, herons, and pelicans). The importance of dredged material islands to nesting species tends to decrease as the size increases because larger islands are more likely to support resident predators. However, isolation is more important than size, and so large isolated islands may be very attractive to nesting birds. Refer to WES TR DS-78-18 for specific information regarding island habitat development.
2. Dredged material islands are found in low- to medium-energy sites throughout the United States. Typically, these are sandy islands located next to navigation channels and are characteristic of the Intracoastal Waterway. In recent years, many active dredged material islands have been diked to improve the containment characteristics of the sites.
3. The importance of dredged material islands as nesting habitats for sea and wading birds cannot be overemphasized. In some states (e.g., North Carolina and Texas), most nesting of these colonial species occurs on man-made islands.
4. Island habitat development has the following advantages: It employs traditional disposal techniques, it permits reuse of existing disposal areas, it provides critical nesting habitats, and its management is conductive to subsequent disposal.
5. Island habitat development uses a traditional disposal technique: the confined or unconfined disposal of dredged material in marsh or shallow water or on existing islands. Consequently, unconventional operational problems seldom occur in its implementation.
6. In many coastal areas, the careful selection of island locales and placement will encourage use by colonial nesting birds. Properly applied, island habitat

development is an important wildlife management tool: It can replace habitats lost to other resource priorities, provide new habitats where nesting and roosting sites are limiting factors, or rejuvenate existing disposal islands.

7. Planned disposal on existing dredged material islands is often conducive to their management for wildlife. Nesting is almost always keyed to a specific vegetation successional state, and periodic disposal may be used to retard succession or set it back to a more desirable state. As a practical matter, disposal on existing islands has largely replaced new island development because of opposition to the loss of open-water and bottom habitats. Consequently, habitat development on dredged material islands will frequently be keyed to the disposal on and management of existing islands.
8. Island habitat development has the following disadvantages: It may interrupt hydrologic process, it may destroy open-water or marsh habitats, and it requires careful placement of material and selection of the disposal season to prevent disruption of active nesting.
9. Alteration of the water-energy regime by the placement of barriers such as islands deserves particular attention because it can change the temperature, salinity, circulation patterns, and sedimentation dynamics of the affected body of water. Large-scale projects or projects in particularly sensitive areas may warrant the development of physical, chemical, and biological models of the aquatic system before project implementation.
10. Dredged material islands, by the nature of their location, may reduce the presence of wetlands and/or open water and their associated benthic habitats. This impact will be minimized by careful site selection of disposal on existing sites. Containment behind dikes will lessen the lateral spread of material but will probably adversely affect the value of the island to birds.
11. Disposal on any dredged material island should be immediately preceded by a visit to determine if the site is an active nesting colony. The use of dredged material islands by birds will occur with or without management. When colonies are present, scheduling of subsequent disposal operations and placement of material should be planned to minimize disruption of the disposal operations as well as of the nesting colonies involved. Destruction of the nests of all colonial water birds is a criminal offense punishable by fine and/or imprisonment.

AQUATIC HABITAT DEVELOPMENT

1. Aquatic habitat development refers to the establishment of biological communities on dredged material at or below mean tide. Potential developments include such communities as tidal flats, sea grass meadows, oyster beds, and clam flats. The bottoms of many water bodies could be altered

using dredged material; in many cases, this would simultaneously improve the characteristics of the site for selected species and permit the disposal of significant quantities of material. Planned aquatic habitat development is a relatively new and rapidly moving field; however, with the exception of many unintentional occurrences and several small-scale demonstration projects, this alternative is largely untested.

2. Aquatic development has major advantages: It produces habitats that have high biological production and potential for wide application and can effectively complement other habitats.
3. Aquatic habitats may be highly productive biological units. Sea grass beds are recognized as exceptionally valuable habitat features, providing both food and cover for many fish and shellfish. Oyster beds and clam flats have high recreational and commercial importance. Dredged material disposal projects affecting aquatic communities often incur strong criticism, and in these instances reestablishment of similar communities may be feasible as a mitigation or enhancement technique. In many instances, it will be possible to establish aquatic habitats as part of marsh habitat development.

Reference

USACE (US Army Corps of Engineers). 2015. *Dredging information system.* Accessed February 2, 2016. http://www.navigationdatacenter.us/dredge/drgtype.htm.

Index